# 化妆品英语听说教程
## Cosmetic English
## Listening and Speaking

主编 孙志青 杨 卉 赵筱婧
主审 张婉萍

上海交通大学出版社
SHANGHAI JIAO TONG UNIVERSITY PRESS

**内容提要**

　　《化妆品英语听说教程》采取ESP基于内容的语言教学观，以化妆品专业基础知识为主线，设计英语听说教学活动。本教程采用新形态教材编写方式。全书包括十个单元，涵盖化妆品发展历史、安全监督与行业规范、护肤、清洁、彩妆等不同品类产品、品牌文化与市场发展等主题。教材以化妆品英语听说能力培养为核心，每个单元设置主题专业概念（Key Concepts）、学习目标（Learning Objectives）、听力训练（Listening）、口语讨论（Speaking）、思辨项目研讨（Critical Thinking）、专业术语（Glossary）五大板块。教材主要可供普通高等院校的化妆品专业学生使用，其他院校的相关专业学生、英语学习者、化妆品行业从业人员也可根据实际情况选用。

**图书在版编目（CIP）数据**

　　化妆品英语听说教程 / 孙志青，杨卉，赵筱婧主编
. —上海：上海交通大学出版社，2023.8（2024.6重印）
　　ISBN 978-7-313-29267-4

　　Ⅰ.①化… 　Ⅱ.①孙… 　②杨… 　③赵… 　Ⅲ.①化妆品
—英语—听说教学—教材 　Ⅳ.①TQ658

　　中国国家版本馆CIP数据核字（2023）第150901号

**化妆品英语听说教程**
HUAZHUANGPIN YINGYU TINGSHUO JIAOCHENG

主　　编：孙志青　杨　卉　赵筱婧
出版发行：上海交通大学出版社　　　　　　　地　　址：上海市番禺路951号
邮政编码：200030　　　　　　　　　　　　　电　　话：021-64071208
印　　制：上海新华印刷有限公司　　　　　　经　　销：全国新华书店
开　　本：889mm×1194mm　1/16　　　　　　印　　张：10.5
字　　数：503千字
版　　次：2023年8月第1版　　　　　　　　　印　　次：2024年6月第3次印刷
书　　号：ISBN 978-7-313-29267-4　　　　　电子书号：ISBN 978-7-89424-403-1
定　　价：45.00元

版权所有　侵权必究
告读者：如发现本书有印装质量问题请与印刷厂质量科联系
联系电话：021-56324200

# 前　言

　　《化妆品英语听说教程》从化妆品专业人才的英语语言听说能力培养的实际需求出发,采取 ESP 基于内容的语言教学观(content-based),以化妆品专业基础知识体系为主线和语言载体,设计英语听说教学活动。教材编写团队由大学英语教师和化妆品专业教师组成,专业教师为教材的专业知识体系及选材内容评估提供重要意见,语言教师负责选材的篇章结构与语言质量评估、教材体例与教学活动设计等方面的工作。

　　教材共含十个单元,涵盖化妆品发展历史、安全监督与行业规范、护肤、清洁、彩妆等不同品类产品、品牌文化及市场发展等主题,内容丰富。教材以化妆品听说能力培养为核心,每个单元设置主题专业概念(Key Concepts)、学习目标(Learning Objectives)、听力训练(Listening)、口语讨论(Speaking)、思辨项目研讨(Critical Thinking)、专业术语(Glossary)五大板块。主题专业概念板块梳理每单元学习需要了解的化妆品专业的相关概念,便于学生课前预习;听说训练板块根据主题和选材难度,分为基础训练和能力提升两部分,满足不同能力层次学生的学习需求,并提供相关专业术语的背景信息、词源信息等说明,帮助学习者加深理解;思辨项目研讨板块聚焦化妆品行业前沿焦点或热点话题,并设计项目式探究任务,提升学生语言表达的同时,加深其对化妆品行业了解,提升批判性思维能力;专业术语板块梳理该单元所涉及的专业英语词汇,并配有音标、中文释义,帮助学习者参阅理解。

　　教材采用新形态教材编写方式,兼顾课堂教学与课后自主学习的教学实际需求,学习者扫描页面上的二维码即可获得该部分的音频、参考答案等教材资源。教材选材新颖,内容覆盖面广,语言精练地道,口语话题和活动设计生动有趣,主要供普通高等院校的化妆品专业学生使用,其他院校的相关专业学生、化妆品行业从业人员、英语学习者也可根据实际情况选用。

　　衷心感谢上海应用技术大学国际化妆品学院院长张婉萍教授,以及张倩洁博士、郑时莲博士、蒋汶博士、张冬梅博士五位化妆品专业教师在教材的体例设计、选材编写、书稿审核等方面的无私分享与辛勤付出。

# Contents

# Unit 1 Introduction

**Lesson 1** Cosmetics: Definition and History

## Checklist for Students

### Key Concepts

| | | | |
|---|---|---|---|
| cosmetics | drug | FDC | cosmeceutical |
| inactive ingredient | active ingredient | abrasive | antioxidant |
| chelating agent | color additive | flavoring agent | fragrance |
| moisturizer | pH buffer | preservative | propellant |
| solvent | surfactant | thickener | anti-acne ingredient |
| antidandruff ingredient | | antiperspirant ingredient | |
| skin protectant ingredient | | sunscreen | |

### Learning Objectives
- Describe the brief history of cosmetics.
- Differentiate between cosmetic products and cosmetic-drug products.
- Understand the gray area between cosmetics and drugs.
- List major types of cosmetic ingredients and major active ingredients.

## Before Listening

People may assume that cosmetic products are a recent invention. However, they have been around almost about as long as human beings have existed. It is believed that the history of the earliest use of cosmetics can be traced back to 6,000 years ago to the ancient Egypt, where makeup served as a marker of wealth that was believed to appeal to the gods. The next hundreds of years have witnessed the development of cosmetics which was full of twists and turns, until the business finally grows into a global industry. In the modern time, however, the concept of cosmetics is becoming much more complex. As the border between is getting blurred, people now have a hard time differentiating between cosmetics and some drug-cosmetic combination products. Whether a product is a cosmetic or a drug is usually determined by its intended use. Nevertheless, since some products meet the definitions of both cosmetics and drugs, it is still taxing to see the difference, and factors like ingredients and ways to apply should also be considered.

## Listening

1.1    Section 1

*Section 1    Brief History of Cosmetics*
*Listen to the audio clip and fill in the blanks with the missing information.*

| Ancient Egypt: first archeological evidence of cosmetics | |
| --- | --- |
| 1. What cosmetic products did the royalty and high class in Egypt enjoy? | They enjoyed 1) _____, perfumed oils, eyeliners, 2) _____, castor oil, lipsticks and 3) _____. |
| 2. Why were cosmetics exclusive at that time? | One of the most important causes for that were their 4) _____, which often included 5) _____ that could cause serious illnesses. |
| 3. What was considered to be the most sacred cosmetic product? | It's 6) _____ because it was used in the 7) _____. |

| Second half of 19th century Europe: dawn of cosmetics | |
| --- | --- |
| 1. Why was the production of cosmetics easier and varied? | Because there was the 8) _____ and then great advances in 9) _____. |
| 2. What changes could be found in cosmetic products? | The price was 10) _____ and the ingredients were 11) _____ for health. |
| 3. What were the most famous cosmetic products? | They were rouge red lipstick, 12) _____, and eyeshadow and 13) _____. |

| 1920s: turning point of cosmetics | |
| --- | --- |
| 1. What methods were employed to promote cosmetics? | Photography, cult of film actors and 14) _____ soon brought the fall of traditional Victorian fashion. |
| 2. What inventive products were there during the time? | There were Lip' Gloss by Max Factor, synthetic hair dye and sunscreen by L'Oréal, suntan produets and 15) _____ by Coco Chanel, and others. |

**Notes**

The dramatic black eyeliners, which were an important part of ancient Egyptian makeup, can be seen from both hieroglyphs and pictures of ancient Egyptian beauties such as Queen Nefertiti or Cleopatra. Scientists have discovered that the iconic Egyptian eyeliner may have health benefits. French researchers analyzed samples from 52 kohl containers residing at the Louvre Museum in Paris and found that the cosmetic contained trace amounts of lead salts which can produce nitric oxide when they come into contact with the skin. Nitric oxide stimulates immune cells, helping to prevent infections of the eye and assisting with the treatment of bacterial eye diseases. This suggests that kohl was more than just a beautifying cosmetic and the forefather of sunglasses, but also an important antibacterial ointment.

### Section 2  Definitions of Cosmetics and Drugs
*Listen to the audio clip and fill in the blanks with the missing information.*

1.1   Section 2

## Cosmetics

In the US Food, Drug and Cosmetic Act or FDC Act, cosmetics are defined as "articles intended to be rubbed, poured, sprinkled, or sprayed on, introduced into, or otherwise applied to the human body or any part thereof for 1) _____ _____ without affecting structure or function." A variety of products are included in this definition, such as skin moisturizers, lipsticks, nail polishes, eye and facial makeup products, shampoos, permanent waves, hair coloring products, and deodorants. In fact any material 2) _____ _____ may fall into this category.

## Drugs

Drugs, on the contrast, are defined in the FDC Act as "articles intended for use in the diagnosis, cure, mitigation, treatment, or prevention of disease" and "articles (other than food) intended to 3) _____ of man or other animals". The significant difference between cosmetics and drugs is easily seen.

## Cosmeceuticals

As a combination of "cosmetics" and "pharmaceuticals", the term "cosmeceutical" caught consumers' interests in the 20th century. It was used for prescription-only products then and addressed appearance issues, such as acne. Today, cosmeceutical mainly refers to multifunctional products that can be purchased as cosmetics and that are advertised to offer 4) _____. Yet, the FDA doesn't recognize "cosmeceutical", although it is a frequently used word by skin care professionals and physicians. The FDA states "a product 5) _____ _____; therefore, the term cosmeceutical has no meaning under the law."

**Notes**

Cosmeceuticals are products that have both cosmetic and therapeutic effects. Cosmeceutical is a globally popular industry, with different names in different countries, such as active cosmetics in Europe and functional cosmetics in South Korea. Regulations about this sector are also vague in many countries. Japan is a country that has legally established a sector called "quasi-drug" between medical drugs and cosmetics, thereby giving "cosmeceutical" legal status and special supervision. The Chinese government holds a very cautious attitude to cosmeceutical, however. In January 2019, the National Medical Products Administration (NMPA) clearly stated that cosmetics claims, such as "cosmeceuticals" and "medical skincare products", are illegal. At the end of the same year, the NMPA issued another announcement, stating that cosmetic products are prohibited from using medical terms and words that indicate or imply a medical effect.

### Section 3  The Gray Area Between a Drug and a Cosmetic Product
*Listen to the audio clip and tell if the following statements are true (T) or false (F).*

1. The division between cosmetic products and drugs has always been blurred.
2. Effects like "repair sun damage to the skin" or "repair skin aging" can be achieved by both cosmetics and drugs.
3. One difference between a cosmetic product and a drug lies in the concentration of the active ingredients in the product.
4. Ingredients like α-hydroxy acids have effect on epidermis even in low concentrations.
5. The border between cosmetic treatments and medical operations is clear cut.

1.1    Section 3

### Section 4  Major Ingredient Types in Cosmetics
*Listen to the audio clip and give short answers to the questions below.*

1. What ingredients do cosmetics contain? What about drug-cosmetic products?

_____

_____

1.1    Section 4

2. What are cosmetic ingredients used for?

_____

_____

3. What are the most commonly used types of cosmetic ingredients? Name as many as you can.

_____

_____

4. What are the major active ingredient types? Name as many as you can.

_____

_____

1.1　Further Listening

## Further Listening

### The Chemistry of Cosmetics

*Listen to the audio clip and choose the correct answer to each of the following questions.*

## Words and Expressions

primer *n.* 妆前乳　　　　emulsion *n.* 乳状液　　　　disperse *v.* 分散
emulsifier *n.* 乳化剂　　　droplet *n.* 液滴　　　　　hydrophilic *adj.* 亲水的
lipophilic *adj.* 亲脂的　　　texture *n.* 质地　　　　　emollient *n.* 润肤剂
hydrogel *n.* 水凝胶　　　　titanium *n.* 钛　　　　　　cochineal *n.* 胭脂虫
carmine *adj.* 洋红色,深红色　hydroxyethyl cellulose 羟乙基纤维素

1. Which of the following is NOT true about emulsifier?
    A) It helps small droplets of oil disperse in water.
    B) One end of the emulsifier molecule structure is hydrophilic and the other end is lipophilic.
    C) It is synthetic and can't be found in natural substances.
2. According to the audio, which of the following is NOT a function of the waxes used in cosmetics?
    A) They keep emulsion from separating.
    B) They keep molecules from absorbing water and swelling.
    C) They keep lipsticks from melting easily.
3. What is the function of hydroxyethyl cellulose?
    A) It is used as a sweetener.
    B) It is used as a thickener.
    C) It is used as a pigment.
4. Where does the color "carmine red" of lipsticks come from?
    A) It comes from organic pigment.
    B) It comes from inorganic pigment.
    C) It comes from cochineal bugs.
5. According to the speaker, what are emollients used for?
    A) They are used to soften your skin by preventing the water loss.
    B) They are used to give your cream the texture.
    C) They are used to keep the products free from bacteria and fungi.

## Speaking

### Section 1　Different Names of Cosmetic Products
*Read the paragraph below and interpret it into Chinese.*

While the FDA only defines cosmetics and drugs, consumers and companies often use the terms "personal care product", "decorative care product", "makeup", "color cosmetic", and "toiletries". But what are these terms? The term "toiletries" is often used for products that are used to clean the body, hair, and teeth, for example, a bodywash, shampoo, and toothpaste, respectively. This term is quite often used interchangeably with personal care products. The terms "color cosmetics", "makeup", and "decorative care products" are generally used for products primarily applied by women to make themselves more attractive, for example, a lipstick, mascara, and nail polish. However, it should be kept in mind that these terms do not reflect the legal state of the products, i.e., whether they are cosmetics or drugs. Even a lipstick can be a drug if it contains sunscreens.

### Section 2　Think and Discuss
*Work in a group. Discuss the following questions and share your answers.*

1. What cosmetic or makeup products does your mum use? What did she use twenty years ago? Are there any similarities or differences? Why?
2. What products do you use? Do you use your mum's beauty products or share with her yours? Why or why not?

## Critical Thinking

### Project 1　Beauty Business: a Conversation Between the Old and the New

China has a long history of making and using cosmetic and makeup products. What do you know about the cosmetic production in ancient China? Can it shed some light on modern beauty industry? How could Chinese cosmetic industry gain a foothold in the global market while keeping our own cultural value?

### Project 2　Men's Beauty Boom: Sissynese or Business

Traditionally, beautifying the appearance with makeup products is taken an exclusive "women business", but now the gender disparity in beauty industry is breaking down. Today, many beauty brands tend to market to men to double their customer base. Is it a good idea to promote men's beauty industry? Will it be a step to break the gender stereotypes or just blur the gender differences?

## Glossary

1. α-hydroxy acids α 羟基酸
2. abrasive /əˈbreɪsɪv/ *n.* 磨料，研磨剂
3. anticaries /ˌæntɪˈkeərɪz/ *adj.* 防龋的
4. antioxidant /ˌæntiˈɒksɪdənt/ *n.* 抗氧化剂
5. antiperspirant /ˌæntiˈpɜːspərənt/ *adj.* 止汗的 / *n.* 止汗剂
6. astringent /əˈstrɪndʒənt/ *n.* 收敛水
7. castor oil 蓖麻油
8. cedar oil 雪松油，柏木油
9. chelating agent *n.* 螯合剂
10. cochineal /ˌkɒtʃiˈniːl/ *n.* 胭脂虫
11. cosmetician /ˌkɒzməˈtiʃən/ *n.* 化妆师
12. deodorant /diˈəʊdərənt/ *n.* 祛臭剂，芳香剂
13. dermatology /ˌdɜːməˈtɒlədʒi/ *n.* 皮肤医学
14. emulsifier /ɪˈmʌlsɪfaɪə(r)/ *n.* 乳化剂
15. emulsion /ɪˈmʌlʃn/ *n.* 乳状液
16. epidermis /epiˈdɜːmis/ *n.* 上皮，表皮
17. fragrance /ˈfreɪgrəns/ *n.* 芳香
18. hair dye 染发剂
19. hydrogel /ˈhaɪdrəˌdʒel/ *n.* 水凝胶
20. hydrophilic /ˌhaɪdrəˈfɪlɪk/ *adj.* 亲水的
21. hydroxyethyl cellulose 羟乙基纤维素
22. lead /led/ *n.* 铅
23. lip gloss 唇彩
24. lipophilic /ˌlɪpəʊˈfɪlɪk/ *adj.* 亲脂的
25. lipstick /ˈlɪpstɪk/ *n.* 口红
26. moisturizer /ˈmɔɪstʃəraɪzə(r)/ *n.* 保湿霜
27. nail polish 指甲油
28. OTC drug 非处方药
29. pharmaceutical /ˌfɑːməˈsuːtɪkl/ *adj.* 制药的
30. plasticizer /ˈplæstɪsaɪzə/ *n.* 塑化剂
31. propellant /prəˈpelənt/ *n.* 推进剂
32. remedy /ˈremədi/ *n.* 疗法
33. rouge /ruːʒ/ *n.* 胭脂
34. shampoo /ʃæmˈpuː/ *n.* 香波
35. solvent /ˈsɒlvənt/ *n.* 溶剂
36. sunscreen /ˈsʌnskriːn/ *n.* 防晒霜
37. surfactant /sɜːˈfæktənt/ *n.* 表面活性剂
38. sweetener /ˈswiːtənə/ *n.* 甜味剂
39. synthetic /sɪnˈθetɪk/ *adj.* 合成的
40. titanium /tɪˈteɪniəm/ *n.* 钛
41. therapeutic /ˌθerəˈpjuːtɪk/ *adj.* 治疗的 / *n.* 疗法
42. thickener /ˈθɪkənə/ *n.* 增稠剂
43. toiletry /ˈtɔɪlətri/ *n.* 洗护用品
44. zinc /zɪŋk/ *n.* 锌

1.1　Keys and Scripts

## Lesson 2  Cosmetics: Brands and Marketing

### Checklist for Students

**Key Concepts**

| | | | |
|---|---|---|---|
| beauty industry | TV advertising | marketing | consumer awareness |
| department store | perfumery | milk | cold cream |
| international market | megabrand | local identities | Estée Lauder |
| François Coty | L'Oréal Paris | Elizabeth Arden | Helena |

**Learning Objectives**
- Learn how cosmetics marketing strategies develop along time.
- Understand the idea of megabrands.
- List top cosmetics brands and companies.
- Learn some basics of cosmetics industry in China.

### Before Listening

The modern cosmetic industry has witnessed the rise of some notable brands. Despite the humble beginning, many of them grow rapidly and have expanded their business beyond the native land, thanks to the brilliant executive and marketing strategies employed. One key to their success is the wise selection of the best means to promote their products. From the early newspapers and fashion magazines to TV commercials, their product promotion is always in line with the evolution of media. New time brings new challenges. Today, as more and more consumers flocking to online shopping sites and use the internet for product research, many leading brands follow suit and adopt digital marketing strategies to level up their beauty business, which sheds new light on the future development of international beauty market.

### Listening

*Section 1  The Rise of Skin Care outside the Home*
*Listen to the audio clip and fill in the blanks with the missing information.*

1.2   Section 1

**"Milks" and "Cold creams"**

Milks were made by 1) _____, such as roses, and mixing with

water and intended to 2) _____ the face. Cold creams were made from
3) _____ and used to 4) _____ .
The first widely sold commercial creams were often created by 5) _____ .
Theron T. Pond, a pharmacist in New York state, was among the earliest entrepreneurs in this tradition.
He developed a product from the bark of witch hazel to relieve cuts and burns in 1846.

## Vaseline

Robert Chesebrough once saw a worker clean oil rig to collect residue which was used for
6) _____ . Becoming interested in the supposed healing qualities of
petroleum derivatives, he created the first "petroleum jelly" which was 7) _____
_____ . Labelled as Vaseline, the product was sold widely after Chesebrough succeeded in
getting 8) _____ , including that of the leading British medical
journal *The Lancet*.

## The growth of a mass market

The growth of a mass market for these and other creams was facilitated by major new
developments in media and advertising. Advertising for brand-name goods proliferated in 9) _____
_____ given away by retailers. The sellers of creams also placed their advertisements in mass-
circulation 10) _____ , which had started appearing in the late
eighteenth century and were widespread by the late nineteenth.

---

### Notes

The main ingredient in Vaseline is petroleum jelly, which is a combination of mineral oils,
waxes, and hydrocarbon. While other products might feature petroleum jelly, only the
products with the blue seal feature actual Vaseline. In the early 1900s, an American girl named
Maybell made a mixture of Vaseline, ash and coal dust to enhance her eyelashes and eyebrows
that were burnt in a kitchen accident. Then her brother Thomas Lyle Williams improved this
homemade formula and launched the first commercially available mascara in the US He later
named the product "Maybelline", a combination of his sister's name and Vaseline, and an
iconic American brand was born.

---

*Section 2    Beauty and TV Advertising*
*Listen to the audio clip and fill in the blanks with the missing information.*

1.2    Section 2

Television's impact on advertising was transformational. 1) _____
_____ . Many brands managed to work their way into people's homes, because
companies which sold mass-marketed consumer goods understood how to make good use
of the media.

### Gillette

Gillette held the exclusive television rights to baseball's World Series between 1950 and 1959. In order to reach the female consumers of the Toni brand of hair care, Gillette 2) _____.

### Charles Revson

Charles Revson sponsored *The $64,000 Dollar Question* game show, which offered the largest amount of cash ever awarded by a radio or television show to a contestant, who was asked increasingly difficult questions on a subject of their choice. During the program, an actress pitched Revlon's products. The company 3) _____.

Television also increased the potential for manipulative advertising. Successfully used, it could 4) _____ in less than a decade. The need for big advertising budgets encouraged a premium on scale rather than creativity in itself.

Meanwhile 5) _____, for whom success in America now rested on levels of advertising expenditure that were far beyond those needed in their domestic business.

### *Section 3　Luxury in Flux: Fighting for Market after WWII*
*Listen to the audio clip and give short answers to the questions below.*

1.2　Section 3

1. How much did Guerlain sell in the US by 1947? How did it achieve this?

_____

2. What did newer American brands need to do to get into international markets?

_____

3. What did Estée Lauder have to do to seek access to European markets?

_____

4. What happened after Estée Lauder spilled her Youth Dew on the floor of Lafayette?

_____

> **Notes**
>
> Estée Lauder was the cofounder of the company that bears her name. As a visionary businesswoman and iconic American entrepreneur, she had many extraordinary achievements during her beauty career, one of which is to develop her company, Estée Lauder, Inc., into one of the largest fragrance and cosmetics companies in the world. In her autobiography, *Estée: A Success Story*, she described some of her basic marketing strategies: opening the Estée Lauder counter at each new store in person, offering free promotional items, and remaining personally involved with the company.

1.2   Section 4

### Section 4   Megabrands and Local Identities

*Listen to the audio clip and choose the correct answer to each of the following questions.*

1. What is the major difficulty of international brands entering local markets?
   A) The lack of product variety.
   B) The government's protection of local brands.
   C) The cultural differences in preferences of local consumers.

2. What's the disadvantage of developing multiple brands?
   A) There was not enough money invested in R&D of new products.
   B) Each brand hadn't sufficient budget on advertising.
   C) It's easy for customers to lose brand loyalty.

3. Which of the following is the way that companies create "megabrands"?
   A) To develop the existing categories.
   B) To acquire categories or segments from other companies.
   C) Both A and B.

4. Why was there a particular desire to develop skin care business?
   A) Because it's a prime concern of both the Western and the Asian consumers.
   B) Because it's more profitable than the business of color cosmetics.
   C) Because it's easier for skin care business to enter local markets.

5. Which of the following is NOT true about the megabrands?
   A) Dove was extended from toilet soap to body care products and color cosmetics.
   B) Nivea was extended into men's toiletries.
   C) Both Maquillage and Tsubaki are megabrands of Shiseido.

## Further Listening

1.2   Further Listening

### China Cosmetics Industry

*Listen to the audio clip and give short answers to the questions below.*

## Words and Expressions

discretionary *adj.* 自由决定的     staple *n.* 主要产品     premiumization *n.* 高端

downturn *n.* 经济衰退     prestige *adj.* 有名望的     masstige *adj.* 大众且奢华的

CAGR Compound Annual Growth Rate 复合年均增长率

Proya 珀莱雅     Sulhwasoo（韩国化妆品品牌）雪花秀

Amorepacific（韩国企业集团）爱茉莉太平洋     Whoo（韩国化妆品品牌）后

LG H&H（韩国化妆品及日化集团）LG生活健康

1. Why is cosmetic a unique category with luxury elements? Why is it also a staple category?

   _____

2. What advantage do cosmetics have compared to other staple categories in China?

   _____

3. What do the numbers refer to?
   1) $79 billion _____
   2) $33 billion _____
   3) $145 billion _____

4. Why is it believed that more than half of overall cosmetics sales will occur online?

   _____

5. Why do Japanese and Korean companies shift their focus to premium brands in China?

   _____

6. What does the speaker want to illustrate with the example of Li Ning?

   _____

## Speaking

### Section 1　Arden and Helena
*Read the paragraph below and interpret it into Chinese.*

The beauty business flourished after the First World War. By the 1920s, Elizabeth Arden and her competitors had succeeded in creating a successful business at the luxury end of the market. Elizabeth Arden's early preeminence in salons was challenged by Helena Rubinstein, who began manufacturing her own products in 1917, a move followed by Elizabeth Arden a year later. These heavily advertised brands were taken across the nation over the following decade. By 1925 Elizabeth Arden was manufacturing 75 individual products, owned salons in nine cities, and distributed in the most prestigious department stores in the United States. As their products were premium-priced, they emphasized how they helped maturing women stay young as they got older. Arden in particular promised women they could join high society if they used her products. Rubinstein also emphasized the association between her products and opulence, taking the opportunity to wear lavish jewelry on public occasions, whilst also regularly being photographed wearing a white laboratory coat to emphasize her commitment to the "science of beauty".

### Section 2　Think and Discuss
*Work in a group. Discuss the following questions and share your answers.*

1. How do you learn about the information of skincare and makeup products? Which way do you think is the most effective or attractive one?
2. What do you know about Chinese local cosmetics brands? Do you use these products? How do you like them?

## Critical Thinking

***Project 1 Promote Your Beauty Brand to Chinese Millennials***

The millennials in China are notably more affluent than their predecessors and more willing to spend money on consumer goods like cosmetics. Also, they differ from the old generations in aesthetic tastes and perception of idealized lifestyles. What could local Chinese beauty brands do to attract the young generation and create customer loyalty? Choose one product or one portfolio of products of a Chinese local brand and design a campaign to sell the product/products to the millennials.

***Project 2 Are the Luxury Cosmetics Worth the Splurge?***

These years see an increasing passion of beauty lovers for prestige cosmetics, which makes the luxury cosmetic market prevail. Beauty products with fancy packaging and elevated ingredients, though sometimes excessively pricey, can still attract customers to spend an arm and a leg on them. Are these luxury cosmetics worth the money? Why would people like to spend such large amount of money on beauty products? Please do research on one product or one portfolio of products to find the answers.

## Glossary

1. Amorepacific 爱茉莉太平洋
2. cold cream 冷霜
3. derivative /dɪˈrɪvətɪv/ *n.* 衍生物，派生物
4. Dove 多芬
5. Elizabeth Arden 伊丽莎白雅顿
6. Estée Lauder 雅诗兰黛
7. François Coty 科蒂
8. Galeries Lafayette 老佛爷
9. Gillette 吉列
10. Guerlain 娇兰
11. Helena Rubinstein 赫莲娜
12. LG H&H LG生活健康
13. Maquillage 心机彩妆
14. Nivea 妮维雅
15. petroleum jelly 矿油
16. pharmacist /ˈfɑːməsɪst/ *n.* 药剂师

17. Proya 珀莱雅
18. Shiseido 资生堂
19. Sulhwasoo 雪花秀
20. Tsubaki 丝蓓绮
21. Unilever 联合利华
22. Vaseline /ˈvæsəliːn/ *n.* 凡士林
23. Whoo 后
24. witch hazel 金缕梅
25. Youth Dew 青春朝露（雅诗兰黛旗下香氛沐浴油）

1.2 Keys and Scripts

## Unit 2  Regulation and Legislation

**Cosmetics: Organizations and Laws**

**Checklist for Students**

**Key Concepts**

Food and Drug Administration (FDA)          Good Manufacturing Practices (GMP)
Federal Food, Drug and Cosmetic Act (FD&C Act)  Poison Prevention Packaging Act
Cosmetic Ingredient Review (CIR)             Fair Packaging and Labeling Act
European Commission (EC)                     Cosmetic Product Safety Report (CPSR)
Voluntary Cosmetic Registration Program (VCRP)
International Nomenclature Cosmetic Ingredient (INCI)
EU Cosmetic Products Regulation    manufacturing    labelling    packaging

**Learning Objectives**

● Know the major governmental and independent organizations in cosmetic industry.
● Understand current rules and regulations for cosmetics.
● Know how manufacturing facilities and products are required to register.
● Know the labeling requirements in the US and EU.

**Before Listening**

It is required by the law that cosmetic products must be properly manufactured with safe ingredients and appropriately boxed and labeled before finding their ways into the market. Government agencies such as the FDA hold authority to regulate the industry under laws like the *Federal Food, Drug, and Cosmetic Act* (or *FD&C Act*) and the *Fair Packaging and Labeling Act*. Companies are also encouraged to follow regulatory standards like the *Good Manufacturing Practices* and participate in the FDA's Voluntary Cosmetic Registration Program (VCRP). All these agencies and regulations work together to protect consumers and prevent the marketing of adulterated or misbranded cosmetics.

## Listening

2.1   Section 1

### Section 1   Responsibilities of the FDA

*Listen to the audio clip and fill in the blanks with the missing information.*

The FDA is the main consumer protection agency of the US government. Its origins can be traced back to about the 1850s. Its modern regulatory function began in 1906 with the promulgation of the *Pure Food and Drugs Act*, which was the predecessor of the *FD&C Act*.

### Public health

The FDA has a number of responsibilities in protecting the public health in the US:

The FDA assures safety, efficacy, and security of human and veterinary drugs, 1) _____ _____, medical devices, food supply, cosmetics, and 2) _____ _____.

The FDA also has responsibility for regulating the manufacturing, marketing, and distribution of 3) _____ to protect public health and to 4) _____ _____ by the youth.

The FDA is also responsible for advancing public health by helping to speed up 5) _____ _____.

### Cosmetics

The FDA is responsible for assuring that cosmetics are safe and properly labeled:

It works with various groups and organizations to ensure the safety of cosmetic products and 6) _____ of these products.

The FDA's goals with respect to international harmonization:

to facilitate 7) _____ and promote mutual understanding; to facilitate exchange of 8) _____ with foreign governments; to review and accept 9) _____, and to enforce programs of other countries.

### OTC drug-cosmetic products

The FDA is responsible for safety and labeling as well as efficacy of product:

It inspects the 10) _____ of to cosmetic and OTC drug-cosmetic verify that all procedures performed meet the regulations.

It also works with the US Customs and inspects imported products at the border to ensure that they meet US regulations.

2.1　Section 2

### Section 2　The Importance of the GMPs

*Listen to the audio clip and tell if the following statements are true (T) or false (F).*

1. In the US, cosmetic products must be manufactured in accordance with the GMPs, but drug-cosmetic products don't have to.
2. In the EU, all cosmetic products placed on the market must be manufactured according to the Cosmetic GMP requirements.
3. Not many companies outside the EU follow the GMPs because they don't have the need to enter the EU markets.
4. Following the GMPs is not required in the US because risks can be eliminated through the testing of the final products.
5. Following the GMPs could help the manufacturers to save cost.
6. The GMP Guidelines only state the minimum requirements, and manufacturers can meet them in a flexible way.

**Notes**

ISO is the short name for International Organization for Standardization. Formed in 1946 in London, the nongovernmental organization is the world's largest developer of voluntary international standards. It develops and publishes a wide range of proprietary, industrial, and commercial standards and is comprised of representatives from various national standards organizations. The ISO standards are developed through global consensus, and their purpose is to help break down the barriers to international trade. The organization's abbreviated name — ISO is derived from the ancient Greek word ísos, which means "equal".

2.1　Section 3

### Section 3　Registration of Manufacturing Facilities and Products in the US and the EU

*Listen to the audio clip and complete the following sentences.*

1. In the US, cosmetic manufacturers must register _____ with the FDA.
   A) the manufacturing establishments　　　B) ingredients of the products
   C) neither of the above
2. VCRP does NOT apply to _____.
   A) a shampoo you purchase in the supermarket
   B) a sample of serum you get from a beauty salon
   C) a YSL lipstick you get from your friend as a gift
3. The initial part of VCRP does NOT include _____.
   A) establishment registration　　　B) ingredient safety assessment
   C) product notification
4. The EU Products Regulation does NOT _____.
   A) issue marketing authorization　　　B) monitor the labeling of cosmetic products
   C) provide information on ingredient safety

5. In the EU, _____.

   A) cosmetic products must be registered

   B) camplete premarket notification must be applied

   C) only CMR substances need to be notified

---

**Notes**

Cosmetics are not subject to the FDA premarket approval. Therefore, the VCRP is just a reporting system but not an approval program or a promotional tool. Participation in the VCRP does not mean that the FDA has approved the firm or its products or that a product is a cosmetic as defined in the *FD&C Act*. Using the fact of participation in labeling or advertising is misleading and against the law.

---

### *Section 4    Cosmetics Packaging and Labeling*
*Listen to the audio clip and give short answers to the questions below.*

2.1   Section 4

1. What documents regulate labeling and packaging of cosmetic products?

_____

2. When will a cosmetic be considered "misbranded"? Write down at least three situations.

_____

3. What is the difference of labeling requirements between the US and the EU?

_____

4. What symbols are used respectively to indicate products which have a minimum durability more or less than 30 months?

_____

**Further Listening**

2.1   Further Listening

### *Regulatory Process for Selling Your Cosmetic Products in the EU and the UK*
*Listen to the audio clip and fill in the blanks with the missing information.*

## Words and Expressions

designated *adj.* 指定的，标明的

stability *n.* 稳定性

nominal quantity *n.* 标称量

specification *n.* 规格，规范

compatibility *n.* 共存，兼容性

portal *n.* 门户网站

| Responsible person |
| --- |
| 1) _____<br>Importer<br>Distributor<br>Third designated person/company |
| **2)** _____ |
| Product formula<br>Specifications on the raw materials<br>3) _____ specifications<br>Stability and compatibility tests<br>Specifications of the 4) _____<br>Cosmetic product safety assessment |
| **Labeling** |
| 5) _____<br>List of ingredients<br>Precautions for use<br>Expiry date<br>Nominal quantity<br>6) _____<br>RP name and address<br>7) _____ |
| **8)** _____ |
| 9) _____ for the EU<br>10) _____ portal for the UK |

## Speaking

### Section 1　Beauty Industry Is Guilty of Misleading Ads

*Read the paragraph below and interpret it into Chinese.*

According to the latest statistics by the Advertising Standards Authority of Singapore (ASAS), while there was an overall decline in the total number of feedbacks on advertising advice, queries and complaints, the volume of feedbacks on the beauty industry has risen to an all-time high. ASAS has seen an increasing number of companies advertising new types of technology of which effects could not be scientifically substantiated. Beauty brands, being aggressive in their advertising, have often used claims that are misleading, unclear or lack substantiation in a bid to lure customers away from their competitors. Meanwhile, the growing number could also be the result of the increasing demand that consumers have for beauty services and products, as compared to in the past.

### Section 2    Think and Discuss
*Work in a group. Discuss the following questions and share your answers.*

1. What do you know about the laws and regulations in China to supervise the registration, labeling, and efficacy claim evaluation of cosmetics? Are they different from those in the US or the EU?
2. What's your plan for your future career in the beauty industry? What ethics and social responsibility are there in the industry? What norms should you follow?

## Critical Thinking

### Project 1    Cosmetic Claims

Claims are statements found on product labels or in TV ads that indicate the expected positive effects of a product. *Standards for Cosmetic Efficacy Claim Evaluation* came into force in 2021 in China. It requires that cosmetic products with efficacy claims shall be supported by an efficacy evaluation. Companies that make exaggerated and misleading claims will face penalties. What claims are illegal and forbidden according to the *Standards*? Why? Support your points with examples and detailed information.

### Project 2    Cosmetic Industry and Environment

The cosmetics industry places a strong emphasis on improving the environmental sustainability of its activities and products. However, even the products that we think to be all natural, when being produced, can still have devastating effects on the environment. What impact does the beauty industry have on the environment? How can the cosmetics industry improve its environmental impact? Find examples to support your points.

## Glossary

1. batch /bætʃ/ *n.* 一批
2. carcinogenic /ˌkɑːsɪnəˈdʒenɪk/ *adj.* 致癌的
3. compatibility /kəmˌpætəˈbɪləti/ *n.* 共存，兼容性
4. efficacy /ˈefɪkəsi/ *n.* 功效，效力
5. mutagenic /ˌmjuːtəˈdʒenɪk/ *adj.* 诱变的，导致突变的

6. nominal quantity 标称量
7. stability /stəˈbɪləti/ *n.* 稳定性
8. Cosmetic Ingredient Review (CIR) 化妆品成分审查
9. Cosmetic Product Safety Report (CPSR) 欧盟化妆品安全报告
10. Federal Food, Drug and Cosmetic Act

(FD&C Act) 美国联邦食品、药品及化妆品法案

11. Food and Drug Administration (FDA) 美国食品药品监督管理局

12. European Commission (EC) 欧盟委员会

13. Fair Packaging and Labeling Act 公平包装与标签法

14. Good Manufacturing Practices (GMPs) 良好生产规范

15. International Nomenclature Cosmetic Ingredient (INCI) 国际化妆品原料

16. Poison Prevention Packaging Act 危险物品包装法案

17. Voluntary Cosmetic Registration Program (VCRP) 化妆品自愿注册计划

18. CPNP cosmetic product notification portal 欧盟化妆品通报

19. PAO period after opening 开封后保存期限

20. PDP principle display penal 主要展示面

21. PIF product information file 产品信息文档

22. RP responsible person 负责人

23. SCPN submit cosmetic product notification 英国化妆品通报

2.1 Keys and Scripts

## Lesson 2　Cosmetics: Safety and Regulations

### Checklist for Students

#### Key Concepts

| | | |
|---|---|---|
| ingredient | safety testing | Hexachlorophene |
| Bithionol | Personal Care Product Council (PCPC) | |
| Consumer Commitment Code (CCC) | adulterated cosmetic | |
| Mercury Compound | Chlorofluorocarbon Propellant | |

#### Learning Objectives

● Understand some basic principles of safety regulations on cosmetics.

● Name the authorities that regulate cosmetics.

● Explain how safety of cosmetics is established in the US.

● List some major prohibited cosmetic ingredients and explain why they are harmful to the skin.

## Before Listening

In the United States, cosmetic products and ingredients do not need the FDA approval before they find their way to the market. However, safety of cosmetics is regulated and monitored by the agent in the collaboration with nongovernmental organizations like the CIR. Under the law, companies and individuals that manufacture or market cosmetics are legally responsible for the safety of their products. Great flexibility is enjoyed by manufacturers to achieve product safety, through preparing insanitary manufacturing establishments, selecting safe ingredients and having any testing that is appropriate and effective. The FDA can take action against the manufacturer of a cosmetic on the market if there is reliable information to show that a cosmetic does not meet the legal requirement for safety.

## Listening

2.2　Section 1

### Section 1　*Are Harmful Ingredients Allowed in Cosmetics?*
*Listen to the audio clip and fill in the blanks with the missing information.*

No matter whether there is a regulation to prohibit the use of a specific ingredient in cosmetics, a cosmetic product is considered adulterated if it contains anything that 1) _____ when consumers use it according to directions on the label, or in the 2) _____ way.

This goes for all the cosmetics except the coal-tar hair dyes. So long as a coal-tar hair dye has a 3) _____ on the label and 4) _____ _____, it is not against the law even if it is or contains a poisonous or deleterious ingredient that may make it harmful to consumers, and the FDA cannot take action against it. The caution statement reads as follows:

> *Caution: This product contains ingredients which may cause 5) _____*
> *on certain individuals and a 6) _____ according to accompanying*
> *directions should first be made. This product must not be used for dyeing the 7) _____*
> *_____; to do it may cause blindness.*

Some ingredients are generally recognized as safe under the conditions of intended use, but safety issues may also arise on certain occasions. For example, some ingredients may be safe in products such as cleansers that we 8) _____ immediately, but not in products that we leave on the skin for hours. Similarly, ingredients that are safe for use on the 9) _____ may be unsafe when used on the skin or near the eyes. That is to say a same product may be safe when used correctly and unsafe when people use it in the wrong way. That is also the reason why there must be directions for use or warning statements to make sure people use the products safely.

Different from food additives, cosmetic products and ingredients do not need the FDA approval before they go on the market. Under US law, 10) _____ lies with cosmetic manufacturers who are obliged to make sure their products are safe and properly labelled. The FDA can and does take action against cosmetics on the market that do not comply with the law.

## Section 2    Cosmetic Ingredient Review Panel

*Listen to the audio clip and tell if the following statements are true (T) or false (F).*

2.2    Section 2

1. The CIR is a subsidiary penal of the PCPC and is financially supported by it.
2. The CIR only examines cosmetic ingredients related to some issues which occurred in the past.
3. Representatives from the FDA and the PCPC are voting members of the CIR.
4. Reports provided by manufacturers through the VCRP is a source of the CIR.
5. The FDA typically takes action on the findings provided by the CIR.

---

### Notes

The Scientific Committee on Consumer Safety or the SCCS is a scientific advisory board of the EC. Similar to the US CIR, it provides opinions on health and safety risks of non-food consumer products like cosmetic products and their ingredients, and services like tattooing and artificial sun tanning. It reviews particular ingredients for which a concern has been expressed for human health and provides advice on the safety of finished products as well. Findings of their risk assessment are published on the EU's website.

---

## Section 3    Testing and Product Safety

*Listen to the audio clip and choose the correct answer to each of the following questions.*

2.2    Section 3

1. What does the speaker say about animal test?
   A) Manufacturers are required by FD&C Act to test cosmetics on animals.
   B) Animal testing is strictly forbidden in the US.
   C) Manufacturers can employ animal testing if they find it necessary in the substantiation of the product safety.
2. What can be learnt about the Consumer Commitment Code?
   A) Its purpose is to help companies ensure safety of their products.
   B) It is initiated by the CIR.
   C) It is a code that all manufacturers must follow.
3. Which of the following is NOT true about the FDA?
   A) It encourages companies to report product safety through the VCRP.
   B) It buys and analyzes cosmetic products and warns customers about the potential danger.
   C) It cannot conduct inspections of cosmetic firms without noticing in advance.

4. Which of the following is NOT a reason for a product to be adulterated?

   A) It contains poisonous or deleterious substance.    B) It is manufactured under insanitary conditions.

   C) It is marketed without undergoing safety testing.

5. Which of the following statements is NOT true?

   A) Manufacturers must follow certain regulations to ensure their product safety.

   B) Manufacturers usually follow guidelines developed by the industry.

   C) Manufacturers can decide flexibly how to perform safety testing.

---

**Notes**

---

In the UK and across the rest of the EU, testing cosmetic products or their ingredients on animals is banned. This means that it is illegal to sell or market a cosmetic product if animal testing has taken place on the finished cosmetic or its ingredients. However, under special circumstances, some testing on animals may be allowed. This is possible upon consultancy of the European Commission with the SCCS but only when there are serious safety concerns on a widely used cosmetic ingredient.

---

*Section 4    Prohibited and Restricted Cosmetic Ingredients by the FDA*

*Listen to the audio clip and complete the following chart.*

2.2    Section 4

| 1. Mercury Compounds | |
|---|---|
| **Function/Usage** | **Harmful Effects** |
| Used for instant whitening | Cause red patches and chronic acne<br>Trigger the growth of cancerous cell or carcinoma |
| **2. Chlorofluorocarbon Propellants** | |
| **Function/Usage** | **Harmful Effects** |
|  |  |
| **3. Hexachlorophene (Nabac)** | |
| **Function/Usage** | **Harmful Effects** |
|  |  |
| **4. Bithionol** | |
| **Function/Usage** | **Harmful Effects** |
|  |  |

2.2 Further Listening

## Further Listening

### *Are Parabens Safe to Use in Cosmetics?*
*Listen to the audio clip and complete the following sentences.*

### Words and Expressions

parabens *n.* 对羟基苯甲酸酯
hell-bent *adj.* 固执的，不顾一切的
sodium benzoate *n.* 苯甲酸钠

antibiotic *n.* 抗生素
phenoxyethanol *n.* 苯氧乙醇
phthalates *n.* 邻苯二甲酸盐

1. Parabens can _____.
   A) increase the stability of skin
   B) change the pH of cosmetics
   C) increase the shelf life of cosmetics
2. According to the speaker, _____ is not a commonly used paraben.
   A) isopropylparaben          B) methylparaben          C) butylparaben
3. People worry about parabens because research shows that _____.
   A) they may cause mutation in humans and animals
   B) they may interfere with body's natural hormones
   C) they are super toxic to the environment
4. The speaker believes _____.
   A) everyone should purchase paraben-free products
   B) more research is needed to find how parabens affect our health
   C) parabens are more toxic than ingredients like colorants or fragrances
5. According to the audio, _____.
   A) parabens have been prohibited in the EU
   B) the FDA has strict regulation on the use of parabens
   C) parabens are favored by beauty companies because they are comparatively cheap

## Speaking

### *Section 1   Recalls*
*Read the paragraph below and interpret it into Chinese.*

Cosmetics Recall is a voluntary action taken by manufacturers and distributors as their responsibility to protect the public health and well-being from products that present a risk of injury or gross deception or are otherwise defective. Recalls of cosmetics and OTC drug-cosmetic products may be conducted on a firm's own initiative or on an FDA request. Under the *FD&C Act*, the FDA has no authority to order a mandatory recall of a cosmetic product; this process is based on a voluntary action. However, the FDA has roles in facilitating recalls and can take authoritative legal actions against manufacturers that persist in marketing a defective product.

In the EU, product recall is the responsibility of the responsible person assigned by the manufacturer. However, the Regulation states that the competent authority can also withdraw or recall a product from the market if immediate action is necessary or if the responsible person does not take all appropriate measures within the set time limit.

### Section 2    Think and Discuss
*Work in a group. Discuss the following questions and share your answers.*

1. Is animal testing essential to the security evaluation of cosmetics or should it be eliminated? Is it possible to develop safe cosmetic products without having animals suffering in research laboratories?
2. Are there any beauty products that you throw away before you finish them and will never ever buy again? What's wrong with them? Why do you think they have such problems?

## Critical Thinking

### Project 1    Ingredients to Avoid for People with Special Skin Concerns

Although some ingredients are considered safe in ordinary beauty, skin and personal care products, they are thought to pose possible risks to babies or pregnant women, or cause problems for people of certain skin conditions like eczema or acne. What are the unsafe or forbidden ingredients for the special groups of customers to use? What harms do they do to these people? What are your suggestions for them to have a safe skin care routine?

### Project 2    How to Use … Safely?

You may assume that the ingredients of the cosmetic products you find in the market have been tested to ensure that they are safe to use, but that's not the case. A study made by researchers of the University of Notre Dame in 2021 shows that over half of all makeup sold in the US and Canada contains high levels of dangerous toxins, and some certain products are believed to be more toxic than others, such as eye cosmetics, tattoo inks and nail care products etc. If we have to use them, how can we do that in a comparatively safe way? Do a research and give your answers.

## Glossary

1. aerosol /ˈeərəsɒl/ *n.* 喷雾器
2. antibiotic /ˌæntibaɪˈɒtɪk/ *n.* 抗生素
3. antiseptic /ˌæntiˈseptɪk/ *n.* 抗菌的
4. bithionol /ˈbɪθənɒl/ *n.* 硫双二氯酚
5. cancerous /ˈkænsərəs/ *adj.* 癌的，生癌的
6. carcinoma /ˌkɑːsɪˈnəʊmə/ *n.* 癌
7. chlorofluorocarbon /ˌklɔːrəʊˈflʊərəʊkɑːbən/ *n.* 氯氟烃
8. chronic acne 慢性痤疮
9. coal-tar /ˈkəʊl tɑː(r)/ *n.* 煤焦油
10. compound /ˈkɒmpaʊnd/ *n.* 混合物，化合物
11. deleterious /ˌdeləˈtɪəriəs/ *adj.* 有毒的，有害的
12. eyebrow /ˈaɪbraʊ/ *n.* 眉毛
13. eyelash /ˈaɪlæʃ/ *n.* 睫毛
14. fluorescent /fləˈres(ə)nt/ *adj.* （物质）有荧光的 /*n.* 荧光灯
15. hexachlorophene /ˌheksəˈklɔːrəfiːn/ *n.* 六氯酚

16. irritation /ˌɪrɪˈteɪʃ(ə)n/ *n.* （身体某部位的）刺激，疼痛
17. mercury /ˈmɜːkjəri/ *n.* 汞，水银
18. methylene chloride *n.* 二氯甲烷
19. paraben /ˈpærəben/ *n.* 对羟基苯甲酸酯
20. phenoxyethanol *n.* 苯氧乙醇
21. photosensitivity /ˌfəʊtəˌsensɪˈtɪvɪti/ *n.* 光敏性
22. phthalate /ˈθælɪt/ *n.* 邻苯二甲酸盐
23. poisonous /ˈpɔɪzənəs/ *adj.* 有毒的
24. propellant /prəˈpelənt/ *n.* 推进剂
25. sodium benzoate *n.* 苯甲酸钠
26. toxicologist /ˌtɒksɪˈkɒlədʒɪst/ *n.* 毒理学家

2.2 Keys and Scripts

# Unit 3　Skin and Hair

## Lesson 1　Skin Structure and Skin Types

### Checklist for Students

**Key Concepts**

| | | | |
|---|---|---|---|
| epidermis | dermis | hypodermis | stratum basale |
| stratum corneum | stratum lucidum | stratum granulosum | stratum spinosum |
| basal cell layer | prickle cell layer | granular layer | keratohyalin granule |
| keratinocyte | translucent layer | horny layer | papillary dermis |
| reticular dermis | subcutis | UVB | transepidermal water loss (TWEL) |
| hydration state | lipid content | normal skin | dry skin　　oily skin |
| combination skin | sensitive skin | skin bacteria | skin color　　pathogen |
| melanocyte | | | |

**Learning Objectives**
- Name the layers of the human skin.
- Explain the functions of the skin.
- Understand moisture content of normal skin.
- Differentiate between normal, dry, oily, combination, and sensitive skin.

### Before Listening

　　The skin is the largest organ of the body and its first line of defense. It protects our internal organs from the environment with a three-layered system of dermis, epidermis and hypodermis. It also plays an active role in regulating body temperature, maintaining water and electrolyte balance and sensing painful and pleasant stimuli etc. Familiarity with the structure and function of the skin is essential to maintain skin health, build reasonable skincare regimen and select proper cosmetics.

## Listening

3.1  Section 1

### Section 1  Skin Structure

*Listen to the audio clip and fill in the blanks with the missing information.*

### Epidermis

| Latin names | Also known as | |
|---|---|---|
| Stratum basale | 1) _____ | It contains cells which continuously 2) _____ _____. |
| Stratum spinosum | 3) _____ | It is responsible for 4) _____. |
| Stratum granulosum | granular layer | In this layer, cells continue to 5) _____ and keratinocytes 6) _____. |
| Stratum lucidum | 7) _____ | It contains 3−5 rows of densely packed 8) _____ that offer extra protection. |
| Stratum corneum | 9) _____ | In this layer, dead cells continuously 10) _____. |

### Dermis

| Two layers | |
|---|---|
| Papillary dermis | It is composed of tissues made up of 11) _____. |
| Reticular dermis | It contains cells that synthesize collagen and elastin which give the skin 12) _____ and 13) _____ after being stretched. |

### Hypodermis

It is known as the 14) _____ and functions as 15) _____.

> **Notes**
>
> Collagen and elastin are proteins in the form of fibers. Collagen fibers give the skin its strength, while elastin is responsible for the skin's elasticity — its ability to "spring back" to its original form after being stretched. If these fibers are damaged as a result of aging, or from excessive, cumulative exposure to the sun, the skin becomes loose and looks thin and wrinkled. In addition, collagen plays an important role in wound healing.

*Section 2   Functions of the Skin*
*Listen to the audio clip and fill in the blanks with the missing information.*

The skin is the largest organ of the body, accounting for about 15% of the total adult body weight. It performs many important functions, such as protecting the body against injury or trauma, regulating the body's temperature, producing Vitamin D, transmitting sensations etc.

## Protection

Only with the protective function of the skin can humans survive in the hostile environment where the temperature and water content vary from time to time and 1) _____ are all round. The skin also prevents water loss of the body. Keratin produced by the keratinocytes in the epidermis is tough and waterproof and prevents water from getting out. This important property helps us to avoid the dehydration in which human bodies may lose great amounts of water to an extent that would threaten life.

## Temperature Regulation

The skin also plays a role in temperature regulation. When the body is cold, vasoconstriction occurs. The blood vessels in the dermis narrow, 2) _____ and causing a person to look pale. In this way the body conserves heat. The blood vessels in the dermis may become wider, which is called vasodilation. This allows more blood to flow through them so as to lose heat if the body is too warm. A person may look more reddish in this situation, which is frequently seen during exercise.

## Production of Vitamin D

Deep in the dermis there is a chemical called 7-dehydrocholesterol. When it is hit by UVB in the sunlight, 3) _____ and it is converted into vitamin D3. The vitamin then passes from the skin into the blood and reaches the various tissues of the body where it exerts its effects. It promotes the absorption of calcium for our bones, without which adults may experience pain and weakness and children can develop rickets.

## Transmission of Sensations

Just by touching, we can tell the texture of things, like whether they are soft, smooth, tough, and hard etc. The sense of touch is vital for humans to interact with the external world. Skin as the organ for touch 4) _____ and is responsible for all the sensations we can feel. The skin has separate receptor for every stimulus which helps to collect all sorts of information about the surroundings.

**Notes**

Although the skin on the lips also has three layers, the stratum corneum on your lips is far thinner than it is anywhere else on the body. The lip skin doesn't have any sebaceous glands either, with saliva as the only source of moisture. These factors make the lip skin particularly sensitive to chemical, physical, and microbial damages. Using lip balms can form a protective barrier on the lips against the harsh environment (wind, cold and UV rays), reducing the risk of having cheilitis and lip cancer.

### Section 3   Transepidermal Water Loss (TEWL)

*Listen to the audio clip and give short answers to the questions below.*

3.1   Section 3

1. What is TEWL and what is the average TEWL of an adult?

_____

2. What are the causes of a high TEWL?

_____

3. How do humectants work in environments respectively?

_____

4. What else can be done to decrease TEWL besides using moisturizing ingredients?

_____

3.1   Section 4

### Section 4   Skin Types

*Listen to the audio clip and write down the features of each skin type in the following chart.*

| Normal skin |
| --- |
|  |
| **Oily skin** |
|  |
| **Dry skin** |
|  |
| **Combination skin** |
|  |

3.1   Further Listening

## Further Listening

***The Colors of Skin: What is Skin Color Determined by?***

*Listen to the audio clip and complete the following sentences.*

### Words and Expressions

melanin *n.* 黑色素                    melanocyte *n.* 黑素细胞          vesicle *n.* 囊泡

melanosome *n.* 黑素体，黑色体   keratinocyte *n.* 角化细胞        eumelanin *n.* 真黑素

pheomelanin *n.* 褐色素           albinism *n.* 白化病                 albino *n.* 白化病者

hemoglobin *n.* 血红蛋白          beta-carotene *n.* β-胡萝卜素   lymph node *n.* 淋巴结

1. Melanocyte cells are found in the _____.
   A) epidermis
   B) dermis
   C) hypodermis
2. People have albinism when _____.
   A) too much eumelanin is produced
   B) too much pheomelanin is produced
   C) no melanin is produced
3. When people are exposed to sunlight or UV radiation, _____.
   A) more melanin is produced to protect the DNA in the nucleus
   B) skin appears reddish because blood vessels increase
   C) it's easy to grow dark age spots
4. _____ is NOT a cause of freckles.
   A) Heredity
   B) Overconsumption of beta-carotene
   C) Exposure to the sun
5. When a bruise heals, the color of it changes from _____ to _____ and then to _____ before
   the skin returns to the normal color.
   A) yellowish; bluish or purple; reddish
   B) reddish; bluish or purple; yellowish
   C) bluish or purple; reddish; yellowish

## Speaking

### Section 1    Skin Thickness
*Read the paragraph below and interpret it into Chinese.*

Skin thickness ranges from 1 to 4 mm. This thickness, and those of each of its layers, varies in different areas of the body.

The epidermis is generally thin. It is particularly so in the skin of the eyelids: approximately 0.1 mm. The epidermis is particularly thick in the soles and palms, where it is approximately 1 mm deep.

The dermis is up to 20 times as thick as the epidermis. It tends to be particularly thick on the back, where it can be approximately 3 to 4 mm.

There is also variability in the thickness of the subcutis skin layer, which tends to be thicker in the thigh and abdominal areas, and particularly thin in the face.

### Section 2    Think and Discuss
*Work in a group. Discuss the following questions and share your answers.*

1. What is your skin type? How do you describe your skin status? Do you have any personal skincare tips?
2. What bad habits will lead to or aggravate skin problems? How will a healthy lifestyle help improve skin health?

## Critical Thinking

### Project 1    Exfoliation

In her book *The Original Beauty Bible*, Paula Begoun claims that every skin type can benefit from exfoliation. Should exfoliation be included in everybody's skin care regimen? What happens when we exfoliate? How to choose proper exfoliants and how to exfoliate without causing skin irritation and inflammation?

### Project 2    Skin on Different Parts of Body

The skin is the body's largest organ, but not all skin is the same. Skin varies in terms of thickness, structure and the way it behaves over different parts of the body. Not all skin gets the same treatment either. Intelligent skincare should reflect the different needs of the skin across the body. Please study how the skin on different parts of our body differs and explain how it should be cared accordingly.

## Glossary

1. acne /ˈækni/ *n.* 粉刺
2. albinism /ˈælbɪnɪzəm/ *n.* 白化病
3. albino /ælˈbiːnəʊ/ *n.* 白化病者
4. allergen /ˈælədʒən/ *n.* 过敏原
5. beta-carotene /ˌbiːtə ˈkærətiːn/ *n.* β-胡萝卜素
6. capillary /kəˈpɪləri/ *n.* 毛细血管
7. dehydration /ˌdiːhaɪˈdreɪʃ(ə)n/ *n.* 脱水
8. dehydrocholesterol /diːhaɪdrəʊkəʊˈlestərɒl/ *n.* 脱氢胆甾醇
9. dermis /ˈdɜːmɪs/ *n.* 真皮
10. eczema /ˈeksmə; ˈeksɪmə/ *n.* 湿疹
11. epidermis /ˌepɪˈdɜːmɪs/ *n.* 上皮，表皮
12. eumelanin /juːˈmelənin/ *n.* 真黑素
13. exfoliation /eksˌfəʊliˈeɪʃn/ *n.* 表皮脱落
14. fungi /ˈfʌŋɡiː/ *n.* 真菌
15. hemoglobin /ˌhiːməˈɡləʊbɪn/ *n.* 血红蛋白
16. humectant /hjuːˈmektənt/ *n.* 湿润剂
17. hypodermis /ˌhaɪpəʊˈdɜːmɪs/ *n.* 下皮，皮下组织
18. keratinocyte /kəˈrætɪnəsait/ *n.* 角质细胞
19. keratohyalin /kerətəˈhaɪəlɪn/ *n.* 透明角质
20. lipid /ˈlɪpɪd/ *n.* 油脂，脂质
21. lymph node *n.* 淋巴结
22. melanin /ˈmelənɪn/ *n.* 黑色素
23. melanocyte /ˈmelənəʊˌsait/ *n.* 黑素细胞
24. melanosome /ˈmelənəʊˌsəʊm/ *n.* 黑素体，黑色体
25. occlusive /əˈkluːsɪv/ *n.* 封闭剂，封包剂
26. pathogen /ˈpæθədʒən/ *n.* 病原体，致病菌
27. papillary /pəˈpɪləri/ *adj.* 乳突的
28. pheomelanin /fiːəʊˈmelænɪn/ *n.* 褐色素
29. photon /ˈfəʊtɒn/ *n.* 光子
30. prickle /ˈprɪk(ə)l/ *n.* 刺
31. receptor /rɪˈseptə(r)/ *n.* 受体，接收器
32. reticular /rɪˈtɪkjʊlə(r)/ *adj.* 网状的
33. scaly /ˈskeɪli/ *adj.* 有鳞屑的
34. sebaceous gland 皮脂腺
35. shed /ʃed/ *v.* 蜕皮，脱毛
36. stratum basale 基底层
37. stratum corneum 角质层
38. stratum granulosum 颗粒层
39. stratum lucidum 透明层
40. stratum spinosum 棘层
41. subcutis /sʌbˈkjuːtɪs/ *n.* 皮下组织
42. transepidermal /trænzˌepɪˈdɜːmɪs/ *adj.* 经表皮的
43. vasoconstriction /ˌveɪzəʊkənˈstrɪkʃn/ *n.* 血管收缩
44. vasodilation /ˌveɪzəʊdaɪˈleɪʃn/ *n.* 血管舒张
45. vesicle /ˈvesɪkl/ *n.* 囊泡

3.1  Keys and Scripts

## Lesson 2  Hair Structure and Hair Types

### Checklist for Students

**Key Concepts**

| | | | | |
|---|---|---|---|---|
| hair follicle | arrector pili | hair bulb | hair papilla | keratinization |
| cuticle | cortex | melanin | medulla | anagen phase |
| catagen phase | telogen phase | exogen phase | lanugo hair | vellus hair |
| terminal hair | hair damage | | | |

**Learning Objectives**
- Understand the structure of hair.
- Describe the life cycle of hair.
- List the major types of hair according to different classifications.
- Understand the factors that cause hair damage.

### Before Listening

Hair is a flexible thin keratin thread with great strength and elasticity. It is present on almost all surfaces of the human skin. For many people the scalp hair is an accessory that can define the characteristic of their appearance. Yet, its role in aesthetics is only part of the function. Hair also collects sweat and protects us from damaging sun rays and particles of debris that could hurt the skin or enter the body. Hair has a complex structure and a regular life cycle. Changes in our hair's look, texture, or thickness can even be signs of underlying health conditions. A good understanding of hair structure, hair types and how it grows and sheds will help us to have healthy and shiny hair.

### Listening

**Section 1  Hair Structure**
*Listen to the audio and make a match between the terms and their definition or function.*

3.2  Section 1

1. Hair follicle

2. Arrector pili

3. Hair papilla

4. Hair bulb

A. major component of hair shaft which decides many mechanical properties of hair

B. an elongated tubular structure that holds the hair

C. the core of the hair shaft

D. a process in which keratin is produced

5. Keratinization

6. Cuticle

7. Cortex

8. Medulla

E. a tiny muscle which can make the hair stand up

F. the base of the root of hair follicle which is in the shape of an onion

G. a piece of tissue with a cluster of blood vessels to support the hair root

H. the outer layer of the hair shaft which is responsible for the texture of it

1. _____    2. _____    3. _____    4. _____

5. _____    6. _____    7. _____    8. _____

### Notes

The average number of hair follicles on the scalp is approximately 100,000. This figure is an average and applies to people with dark hair. The number varies, depending on hereditary factors and the shade of hair. Redheads have relatively less, but thicker, scalp hair (the average is 80,000). People with blond hair have thinner hair, but more of it — approximately 120,000 hair follicles on the scalp. With age there is a gradual loss of hair follicles from the scalp to varying degrees.

### Section 2    Life Cycle of Hair

*Listen to the audio and put the correct description of each phase of life cycle of the hair in the right place.*

3.2    Section 2

| | |
|---|---|
| A. it marks the end of hair growth | B. hair is released and falls out |
| C. hair grows at a rate up to 0.4 mm a day | D. it is a transitional phase |
| E. its span decides how long hair can grow | F. it lasts for three to six months |
| G. most of the scalp hair are in this phase | H. production of pigment stops |
| I. hair is loosely attached to the follicle | J. hair follicle shrinks |
| K. it lasts for two to three weeks | L. dermal papilla stops growing |
| M. it's the longest phase of the life cycle | N. it is the active growing phase |
| O. it is an extension of the telogen phase | P. new hairs grows from the follicle |

1. Anagen Phase: _____

2. Catagen Phase: _____

3. Telogen Phase: _____

4. Exogen Phase: _____

3.2 Section 3

### Section 3 Hair Types

*Listen to the audio clip and fill in the blanks with the missing information.*

| Classification based on thickness | |
|---|---|
| 1) _____ hair | It is more resistant to 2) _____ and 3) _____, such as hair perming. |
| Medium hair | It is usually considered the standard to which other types are compared. |
| 4) _____ hair | It has just two layers: 5) _____. |
| **Classification based on greasiness** | |
| Dry hair | It is not moist enough because of the damaged layers of the hair shaft. Dry hair can be further damaged by chemical treatment such as permanent waving or 6) _____. |
| Oily hair | It is a result of 7) _____. It is associated with problems like hair thinning and 8) _____. |
| **Classification based on shape, size, and color** | |
| Lanugo hair | It is thin, soft hair that covers the human baby inside the womb. |
| Vellus hair | It is usually found on surfaces normally considered hairless and helps in 9) _____ and 10) _____. |
| 11) _____ | Controlled by the 12) _____, the hair is converted from the vellus hair during puberty. |

### Section 4 Hair Damage

*Listen to the audio clip and give short answers to the questions below.*

3.2 Section 4

1. What actions exert tension on the hair and the scalp? Write at least three of them.

   _____

2. What treatments may cause chemical damages to the hair?

   _____

3. What are the features of chemically damaged hair?

   _____

4. What are the environmental factors that may damage the hair?

   _____

5. How does hard water affect our hair?

   _____

3.2   Further
Listening (I)

## Further Listening

### *Why Do We Have Hair in Such Random Places ? (I)*

*Listen to the audio clip and tell if the following statements are true (T) or false (F).*

**Words and Expressions**

mammalian *adj.* 哺乳类的
insulate *v.* 隔热

porcupine *n.* 豪猪
hominin *n.* 古人类

1. Human hair and animal fur are made of the same substances.
2. For the hair layers of mammals, the ground hair is longer than the guard hair.
3. The major function of mammal hair is to prevent the moisture from evaporating from the boy.
4. Hair found in the fossil helps scientists to figure out how humans lost most of the hair along time.
5. Humans have much more sweat glands than chimpanzees.

3.2   Further
Listening (II)

### *Why Do We Have Hair in Such Random Places? (II)*

*Listen to the audio clip and summarize the functions of hair in different body parts.*

**Words and Expressions**

debris *n.* 碎片，残渣
secretion *n.* 分泌物

apocrine gland 顶质分泌腺

1. Hair on the top of head

_____

2. Eyebrows

_____

3. Eyelashes

_____

4. Facial hair like beard

_____

5. Hairs in the armpits and pubic areas

_____

6. Vellus hair (and the follicles from which it grows)

_____

**Notes**

_____

There are two distinct types of temporary hair removal, known as depilation and epilation. Depilatory techniques, include shaving, trimming, using abrasives, using chemical depilatories, and bleaching, remove only part of the hair shaft. Epilatory techniques (tweezing, waxing and treading etc.) remove the entire hair shaft with its root in the dermis and have a longer effect. Though removing unwanted hair has become a part of many people's everyday life, it should be remembered that very frequent and/or inappropriate hair removal can lead to adverse effects and skin reactions.

## Speaking

### Section 1    Hair Planting
_Read the paragraph below and interpret it into Chinese._

Mention hair transplants to someone and it will immediately conjure up images of obvious, unsightly plugs of hair dotting someone's scalp like bad patches of grass on a lawn. Hair transplants from a decade ago did use strips of hair, grafts of scalp, and the results appeared like rows of planted hairs growing from a black plug on the scalp. Or you imagine little plugs dotted over the head appearing like sprouts of chives growing over the head. Fortunately, those days are over, and new techniques in hair transplants create a completely natural look. Today's techniques implant only one to four hairs, with no detectable base in sight. This state-of-the-art procedure is called the follicular-unit grafting technique, a process that relies on microscopic dissection at the back of the head to produce the hair grafts. It is an expensive, complicated procedure, but the results are remarkable and the hair does grow with minimal to no risk of further hair loss or thinning.

### Section 2    Think and Discuss
_Work in a group. Discuss the following questions and share your answers._

1. What is your hair type? How do you describe your hair status? Do you have any personal tips on the care of hair?

2. Do you know any myths of haircare? What are they? Why do you think they are false?

## Critical Thinking

*Project 1    Hair Loss Scams*

It is said that approximately 25% of men begin balding by age 30, and women, though less vulnerable to alopecia, are also affected by pattern baldness. Therefore, it's not difficult to explain why we see so many products (shampoo, conditioner, scalp masks) that claim to have the magic power of generating hair growth or reducing hair loss. Can these products really prevent hair thinning or regrow what we have lost? Or are they just scams? What can we do in our daily life to have wholesome and robust hair?

*Project 2    Hair Removal*

People in modern society are obsessed with getting rid of unwanted body hair, no matter it is the dark hair above the lip or the dense hair growth on the arms or legs. This can be done physically by pulling the hair out of its follicles (plucking/waxing) or using chemical depilatories. Does hair removal do harm to our health? Which ways are safer and more effective? What particular attention should be paid when people remove the body hair?

## Glossary

1. anagen /ˈænədʒ(ə)n/ *n.* 毛发生长初期
2. arrector pili 竖毛肌
3. bleach /bliːtʃ/ *v.* 漂白，使褪色
4. catagen /ˈkætɪdʒen/ *n.* 毛发生长中期，退行期
5. cortex /ˈkɔːteks/ *n.* 皮层
6. cuticle /ˈkjuːtɪkl/ *n.* 角质层
7. exogen /ˈeksədʒen/ *n.* 活跃脱毛期
8. follicle /ˈfɒlɪk(ə)l/ *n.* 毛囊
9. hair bulb 毛球
10. hair papilla 毛乳头
11. hair root 毛根
12. hair shaft 毛干
13. lanugo hair 胎毛
14. medulla /meˈdʌlə/ *n.* 髓质
15. perm /pɜːm/ *v.* 烫发
16. perspiration /ˌpɜːspəˈreɪʃn/ *n.* 排汗
17. pigment /ˈpɪgmənt/ *v.* 给……染色 /*n.* 染料
18. telogen /tɪləʊˈdʒen/ *n.* 毛发生长终期，静止期
19. transverse /ˈtrænzvɜːs/ *adj.* 横向的
20. vellus hair 细毛，柔毛

3.2  Keys and Scripts

## Lesson 3　Skin Problems

### Checklist for Students

**Key Concepts**

| | | | |
|---|---|---|---|
| intrinsic aging | elastin fiber | collagen fiber | thinning of the skin |
| extrinsic aging | photoaging | acne | hair follicle |
| sebaceous gland | open comedone (blackhead) | closed comedone (whitehead) | |

**Learning Objectives**

- Differentiate between intrinsic and extrinsic aging.
- Briefly discuss what changes occur in the skin during aging.
- Name environmental factors that can accelerate the normal aging process.
- Explain how acne occurs.
- Differentiate between blackheads and whiteheads.
- Name the major factors contributing to the development of acne.

### Before Listening

Although most people don't think of it as an organ, skin is, in fact, the biggest organ we have. As such, it requires a lot of our attention and care. Sometimes, even with proper skin care, multiple skin disorders occur on our skin. Innumerable internal and external factors may lead to some of the common skin issues. Fine lines and dark spots will appear as we age, and unhealthy lifestyle and improper skin care regimen may accelerate the process. Anything that irritates, clogs, or inflames the skin can contribute to skin problems like acne, causing redness, swelling and itching. Whatever the cause might be, gaining an insight into skin problems will help us manage and protect our skin in a better way.

### Listening

#### Section 1　Intrinsic Aging
*Listen to the audio clip and fill in the blanks with the missing information.*

3.3　Section 1

　　Intrinsic aging, also known as 1) _____, is the natural aging process which is genetically determined and occurs by the passing of time. It is mainly a result of slow tissue degeneration with changes such as degeneration of the elastin and collagen fibers, thinning of the skin and loss of hydration occurring.

## Degeneration of elastin and collagen fibers

Elastin provides the structure of dermis. Elastin fibers undergo a degenerative process as the body ages and become 2) _____ of poor quality over time. In addition to elastin fibers, 3) _____ also gradually diminishes. When both fibers that give the skin its elasticity are constantly reduced, the skin starts to sag and wrinkles appear.

## Thinning of the skin

With advancing age, there is a gradual thinning of all skin layers, including the 4) _____ _____. As collagen is responsible for the strength of the skin, its decreasing amount leads to reduced strength and weakening and thinning of the dermis. The subcutaneous fatty layer becomes thinner, too. This process of degeneration and waning of tissue is called atrophy. Also, the 5) _____ on the face begin to shrink, causing lines to appear. Lines are more prominent in the areas where muscles contract to 6) _____. That's why we usually see "worry lines" or "laugh lines" on the face.

## Loss of hydration

With increasing age, the skin becomes drier. A primary cause of dry skin is the gradual declining function of the 7) _____. These glands produce a fine lipid layer over the skin surface which serves as a barrier preventing evaporation of water from the skin. With the barrier functions damaged, 8) _____ increases and the skin gets dry.

*Section 2　Extrinsic Aging*
*Listen to the audio clip and complete the following sentences.*

3.3　Section 2

1. _____ is the primary cause of skin aging, which accounts for about _____ of skin problems.
   A) Sun exposure; 90%　　　B) Air pollution; 80%　　　C) Pressure; 60%
2. UVA light _____.
   A) mainly does harm to the dermis
   B) has short waves
   C) could make the skin lose elasticity
3. UVB light _____.
   A) can penetrate through different layers of skin
   B) damages protein and collagen in the dermis
   C) may cause problems like skin cancer
4. In the glycation reaction _____.
   A) advanced glycation end products are formed
   B) byproducts of sugar displace fibers and proteins

C) the formation of dark spots is accelerated

5. According to what you have learnt, a way to slow the aging process is to _____.

   A) smoke e-cigarettes instead of traditional tobacco

   B) do exercise regularly

   C) go vegan

---

**Notes**

Based on the light's wavelength, UV light is divided into three groups: UVC, UVB and UVA. UVC is largely blocked by the ozone layer and has little impact on the skin. UVB penetrates only into the epidermis and is responsible for the redness, sun-burn, DNA damage, hyperpigmentation, and skin cancer. UVA requires a much higher dose to cause sunburn, and for this reason, it was long considered irrelevant to skin damage. However, because it can penetrate deeper into the skin than UVB, today, UVA is thought to play a more substantial role in photoaging.

---

3.3　Section 3

### Section 3　Acne

*Listen to the audio clip and give short answers to the questions below.*

1. How do acnes occur? Fill in the blanks with proper content to complete the mechanism.

   A. _____

   B. The sebum and dead cells cannot get out to the skin surface.

   C. _____

   D. A sebum-rich, oxygen-poor environment is created.

   E. _____

2. What are the major differences between a whitehead and a blackhead?

   _____

3. Among the four types of acne, which are inflammatory ones?

   _____

4. When do the acnes usually first appear? When will they generally disappear?

   _____

3.3　Section 4

### Section 4　Causes of Acne

*Listen to the audio clip and tell if the following statements are true (T), false (F) or not given (NG).*

1. Androgens or male sex hormones can be found in both boys and girls.

2. Acne patients should wash their face more diligently to keep facial hygiene.

3. Deep facial cleansing can reach the deeper layers of the skin and help to eliminate excessive sebum.

4. Some facial cleansers contain irritating ingredients and lead to burning and stinging.

5. Ingredients like silicon, paraben and dyes could induce or promote acne.

6. Acne patients should use fewer or, if possible, no makeup products.

7. There is no direct association between acne and certain foods.

8. Foods with glycemic load could aggravate acne.

## Notes

The effects of sunlight on acne have long been studied. Light therapy is increasingly used in acne treatment, and some studies have shown evidence of sunlight being beneficial to acne patients. The improvement is related to a certain anti-inflammatory effect of the ultraviolet rays. However, exposure to the sun's rays may also cause an excess production of keratin and sebum, both on the skin's surface and in the pores, which may, in turn, cause a relative worsening of the acne. The risk of development of skin cancers must also be taken into account. Therefore, care should be taken in advising therapeutic sun exposure.

## Further Listening

3.3   Further Listening

***Does Pressure Cause Pimples?***
*Listen to the audio clip and give short answers to the questions below.*

### Words and Expressions

zit *n.* 青春痘

sabertoothed tiger 剑齿虎

pituitary gland 脑下垂体

adrenocorticotropic *adj.* 促肾上腺皮质的

neuropeptide *n.* 神经肽

corticotropin releasing hormone 促肾上腺皮质素释放激素

daze *n.* 眩晕，迷茫

hypothalamus *n.* 下丘脑

adrenal gland 肾上腺

cortisol *n.* 皮质醇

testosterone *n.* 睾酮

1. What symptoms could there be if a person is attacked by panic?

_____

2. According to the speaker, when will people internalize stress?

_____

3. How could the stress hormones affect human health?

_____

4. What is the major stress hormone that is involved in making skin cells produce lipids?

_____

5. What could be done to avoid pimples?

_____

## Speaking

### Section 1  Invasive and Non-invasive Ways to Fight Against Aging
_Read the paragraph below and interpret it into Chinese._

Today, there is a wide range of products and procedures for the prevention and/or treatment of aging skin, ranging from non-invasive technologies to invasive methods.

Non-invasive technologies include primary prevention, such as healthy lifestyle, refraining from smoking, and using sunscreens, topical cosmetics and drug products applied directly to the skin, microdermabrasion, certain laser devices, as well as systemic treatments, including hormone replacement therapy.

Invasive technologies include chemical peels, topical injections of chemicals, such as botulinum toxin and various dermal fillers, dermabrasion, various laser devices, as well as corrective surgeries.

A healthy skin barrier is an important protector against dehydration and penetration of various physical and chemical entities and contributes to skin regeneration, elasticity, and smoothness. Therefore, its daily maintenance is an essential part of the anti-aging therapy.

### Section 2  Think and Discuss
_Work in a group. Discuss the following questions and share your answers._

1. How can we slow down intrinsic and extrinsic aging?
2. Do you suffer from acnes? What type is your acne? Can you share some tips to deal with acnes?

## Critical Thinking

### Project 1  Glycation

The food we eat can impact how rapidly our skin ages. Certain foods contribute to what's known as glycation, which not only has impact on our well-being, but also makes us grow old faster. What is glycation? How does glycation impact the skin? What can we do to reduce the effects of glycation process?

*Project 2　Puberty Acne*

Among all the skin problems that trouble teenagers, acne is at the top of the list. The good news is it is a highly treatable condition and can be managed in different ways. Please study the causes of puberty acne and give suggestions on the treatment of it.

## Glossary

1. adrenal gland 肾上腺
2. adrenocorticotropic /əˈdriːnəʊˌkɔːtɪkəʊˈtrɒpɪk/ *adj.* 促肾上腺皮质的
3. atrophy /ˈætrəfi/ *n.* 萎缩症
4. corticotropin releasing hormone 促肾上腺皮质素释放激素
5. cortisol /ˈkɔːtɪsɒl/ *n.* 皮质醇
6. degeneration /dɪˌdʒenəˈreɪʃn/ *n.* 恶化，衰退
7. extrinsic /eksˈtrɪnzɪk/ *adj.* 外在的
8. glycation /glaɪˈkeɪʃən/ *n.* 糖化
9. humidity /hjuːˈmɪdəti/ *n.* 湿度，潮湿
10. hypothalamus /ˌhaɪpəˈθæləməs/ *n.* 下丘脑
11. intrinsic /ɪnˈtrɪnzɪk/ *adj.* 内在的
12. lesion /ˈliːʒn/ *n.* 损害，损伤
13. micro comedo 微小粉刺
14. neuropeptide /ˌnjʊərəʊˈpeptaɪd/ *n.* 神经肽
15. papule /ˈpæpjuːl/ *n.* 丘疹
16. photoaging /fəʊtəʊˈeɪdʒɪŋ/ *n.* 光老化
17. pituitary gland 脑下垂体
18. precancerous /ˌpriːˈkænsərəs/ *adj.* 癌症前期的
19. pustule /ˈpʌstjuːl/ *n.* 脓疱
20. tautness /ˈtɔːtnəs/ *n.* 紧固度
21. testosterone /teˈstɒstərəʊn/ *n.* 睾酮
22. zit /zit/ *n.* 青春痘

3.3　Keys and Scripts

# Unit 4 Skincare Products

## Lesson 1 Toner, Serum and Essence

### Checklist for Students

**Key Concepts**

| | | | | |
|---|---|---|---|---|
| astringent | toner | essence | serum | ampoule |
| alcohol | skin-aggravating ingredient | water menthol | witch hazel | fragrant extract |

**Learning Objectives**
- Name the basic functions of toner.
- List the major ingredients of toner.
- Understand the basics of serum.
- Understand the basics of essence.

### Before Listening

"Skincare product" is a broad concept. Anything used to protect, moisturize or revitalize the skin can be categorized as skincare products. They share many similarities, and the major differences lie in the texture and the concentration of ingredients. Toners, essences and serums have similar textures, while some are more watery and others may be more viscous. Toners are liquid formulas that fulfill different purposes, such as cleansing, adding hydration, or exfoliating. Effective as an in-between step to prepare the skin for other products, toners may contain irritating ingredients that should be avoided. Essences are taken as watered-down serums that tend to have a thinner liquid consistency or water-gel textures. Many essences have similar ingredients that can be found in serums, but the higher water content may allow people with sensitive skin to tolerate active ingredients better than using a serum. Serums contain higher concentrations than other formulations. Serums can come in gel, liquid, or oil form.

## Listening

### Section 1　Toners: General Introduction
*Listen to the audio clip and tell if the following statements are true (T) or false (F).*

4.1　Section 1

1. Toners are mainly used to remove residues of oil, dirt, and makeup.
2. AHAs/BHAs/PHAs are important ingredients in healing toners.
3. Hydrating toners may contain high amount of alcohol and are not good for sensitive skin.
4. Healing toners can be used to reduce redness or get rid of irritation.
5. Toners should be applied with hand.
6. Strong toners should be diluted before use.

**Notes**

Panthenol, which is translated as "泛醇" in Chinese, is a type of hydrating ingredient commonly added in toners. Alcohol compounds used in cosmetics follow the naming principle of adding suffix "-ol" at the end of the English words. Other examples may include methanol (甲醇), ethanol (乙醇), sorbitol (山梨醇), propanediol (丙二醇) and butanediol (丁二醇), etc.

### Section 2　Toners: the Ingredients to Avoid
*Listen to the audio clip and match the descriptions with the ingredients.*

4.1　Section 2

A) is effective in treating black and whiteheads
B) may cause allergic reaction
C) may eliminate sebum
D) may induce peeling
E) is used as anti-acne ingredients
F) is used as antiseptic or antibacterial agents
G) may accelerate oil production
H) may cause contact dermatitis
I) can dry or irritate the skin
J) can disrupt pH balance

1. _____ Alcohol
2. _____ Salicylic acid
3. _____ Benzoyl peroxide
4. _____ Fragrance

**Notes**

Glycolic acid, a type of acidic ingredients, is translated as "乙醇酸" or "甘醇酸" in Chinese. "glycol-" and its other forms such as "gly-" or "glyc-" have the meaning of "sugar or sweet", hence the Chinese equivalent "甘" or "糖". Other examples with the same word root may include glycation (糖化), and glycolysis (糖酵解).

### Section 3    Serum

*Listen to the audio clip and give short answers to the questions below.*

4.1    Section 3

1. What skin problems are serums supposed to deal with?

_____

2. How do hydrating serums work?

_____

3. What ingredients do anti-aging serums contain?

_____

4. What skin problems do free radicals cause?

_____

5. Why is serum a poor match for people with certain skin problems like eczema?

_____

**Notes**

Retinol, a form of Vitamin A, has been a very popular anti-aging ingredient in serums to protect the skin from fine lines and loss of firmness. But it is best applied at night, or together with effective sun protection during the day, because even though retinol itself does not have sun sensitivity, it is unstable under UV light, and the irritation from it will increase the skin's sensitivity to UV light.

### Section 4    Essence

*Listen to the audio clip and fill in the blanks with the missing information.*

4.1    Section 4

**Definition**

An essence is a 1) _____. Some of them are very liquid like water, and the thicker ones have the texture of a very light gel, which is a little more viscous.

## Function

First, essences are meant to deliver highly hydrating ingredients, 2) _____ _____, to deeply nourish and moisturize the skin. Essences will help the serums and moisturizers that are used afterwards to absorb into the skin better. Furthermore, 3) _____. This is important because the use of some harsh soap-based cleansers can strip the skin of its natural oils. Even though we often want to reduce these oils, stripping too much of them can be detrimental to the skin and cause even more oil to be produced. If aging is a major concern, this added boost of moisture can also 4) _____. And for sensitive skin type, it can aid in strengthening the skin barrier.

## Difference between essence and serum

The main difference is in the key purpose of both products. The two main activities of essence are to 5) _____. Serums address a wider range of skin conditions. They have functions like skin hydration, skin brightening, skin resurfacing, treating acne and fading scars, treating signs of aging, and aligning your skin tone. To assist them to tackle such problems, serums generally 6) _____.

## Further Listening

***Essence, Serum and Ampoule: What Are the Differences?***
*Listen to the audio clip and give short answers to the questions below.*

4.1   Further Listening

### Words and Expressions

| | | |
|---|---|---|
| essence *n.* 精华 | serum *n.* 精华（液） | ampoule *n.* 安瓿 |
| concentration *n.* 浓度 | viscosity *n.* 黏度 | dropper *n.* 滴管 |
| consistency *n.* 稠度 | combo skin 混合肌 | mature skin 熟龄肌 |

1. What skincare steps are recommended to beginners?

_____

2. What is the difference between essence and toner?

_____

3. What is the difference between essence and serum?

_____

4. What should you do if you find using ampoule every day too heavy?

_____

5. According to the speaker, what is the general rule in skincare?

_____

6. What skin types are essence, serum and ampoule suitable for respectively?

_____

## Speaking

### *Section 1    Aftershave Preparations*
*Read the paragraph below and interpret it into Chinese.*

Aftershave preparations are made of the same substances as those used in astringents; they also contain water and alcohol. The assumption is that even the low concentration of alcohol present in an aftershave has some antiseptic effect, which is helpful in dealing with tiny cuts or abrasions in the skin (some of them are not even visible or felt) that occur during shaving. Zinc or aluminum salts in the product are said to constrict the skin pores that are dilated following rinsing of the face with warm water. Aftershave lotions give a feeling of freshness and coolness, usually due to the addition of menthol. With regard to aftershave preparations, the only real difference between the various brands is the unique scent of each one. The practical value of astringents is controversial. It has not yet been shown in the medical literature that they indeed have any beneficial effect.

### *Section 2    Think and Discuss*
*Work in a group. Discuss the following questions and share your answers.*

1. How many steps do you have in your skincare routine? What are they? Why do you include or exclude certain steps?
2. Among many expensive skincare products, essence and serum are usually at the top. Why are they so costly? Does higher price mean better quality? How to choose the right essence or serum?

## Critical Thinking

### *Project 1    Evaluation of Different Toners or Essences*

Suppose you and your partner use different toners or essences. What products do you use respectively? Why do you choose these products? Are they effective on your skin? Please make a presentation evaluating the different toners or essences you use by introducing the detailed information of the products, your skin type and the efficacy of the products.

### *Project 2    Witch Hazel*

Today, customers are often advised to avoid toners which contain witch hazel. However, this doesn't affect the natural ingredient's reputation as the solution to endless skin problems. What is witch hazel? Is it good or bad for our skin? How should it be used? Conduct a research on this topic and present your findings.

## Glossary

1. aftershave /ˈɑːftəʃeɪv/ *n.* 须后水
2. AHA/BHA/PHA α羟基酸/β羟基酸/聚羟基脂肪酸
3. ampoule /ˈæmpuːl/ *n.* 安瓿
4. antiseptic /ˌæntiˈseptɪk/ *adj.* 杀菌的
5. astringent /əˈstrɪndʒənt/ *n.* 收敛水
6. azelaic acid 壬二酸
7. dermatitis /ˌdɜːrməˈtaɪtɪs/ *n.* 皮炎
8. dropper /ˈdrɒpə(r)/ *n.* 滴管
9. eczema /ˈeksɪmə/ *n.* 湿疹
10. free radical 自由基
11. glycerin /ˈɡlɪsərɪn/ *n.* 甘油
12. glycolic acid 乙醇酸
13. HA (hyaluronic acid) 透明质酸
14. kojic acid 曲酸

15. lactic acid 乳酸
16. niacinamide /naɪəˈsɪnəmaɪd/ *n.* 烟酰胺
17. peptide /ˈpeptaɪd/ *n.* 肽
18. retinol /ˈretɪˌnɒl/ *n.* 视黄醇
19. salicylic acid 水杨酸
20. SD alcohol 变性酒精
21. serum /ˈsɪərəm/ *n.* 精华液
22. essence /ˈesns/ *n.* 精华，香精，精油
23. toner /ˈtəʊnə(r)/ *n.* 爽肤水

4.1　Keys and Scripts

---

## Lesson 2　Lotion and Cream

## Checklist for Students

### Key Concepts

| | | | |
|---|---|---|---|
| oil-in-water emulsion | water-in-oil emulsion | fatty acid | corneocyte |
| dispersed phase | continuous phase | hyaluronic acid | ceramide |
| lotion | cream | glycerin | occlusive |
| emollient | humectant | | |

**Learning Objectives**

- Understand the differences between lotions and creams.
- Understand the differences between W/O and O/W emulsions.
- List the major lotion ingredients.
- Understand how moisturizers work.

## Before Listening

If you are puzzled by the tubes and tubs labeled as "moisturizing lotion" or "hydrating skin cream" on the shelf in stores like Sephora, you are not alone. As both lotions and creams are designed to hydrate the skin, no wonder that many people have a difficult time distinguishing the differences and choosing the right ones. They are both emulsions made from a combination of water and oil. Yet the ratio of these two ingredients is different in each product type, leading to variations in consistency and greasiness. Choices should be made on the basis of skin types, time or season to use and any particular skincare needs like anti-aging or whitening. No matter which one to choose, using moisturizers regularly should be a part of any good skin care regimen. Proper moisturizing reduces the chance of developing dryness or oiliness, saving you from common skin conditions like acne and helping the skin to stay young.

4.2 Section 1

## Listening

***Section 1    Differences Between Lotions and Creams***
*Listen to the audio clip and complete the following chart.*

| Differences | Lotions | Creams |
|---|---|---|
| water/oil proportion | 1. | 2. |
| texture or consistency | 3. | 4. |
| seasons to use | 5. | 6. |
| time of the day to use | / | 7. |
| skin types | 8. | / |
| containers | 9. | 10. |

4.2 Section 2

***Section 2    Lotion, Cream and Skin Types***
*Listen to the audio clip and give short answers to the questions below.*

1. What factors are to be considered before choosing between creams and lotions?

_____

2. Why are creams better than lotions in cold winter?

_____

3. Should people with oily skin skip moisturizers? Why or why not?

_____

4. What types of alcohol are usually used in lotions? Why?

_____

5. Why are people with acne-prone skin advised to avoid lotions containing alcohol?

_____

*Section 3　Oil-in-Water and Water-in-Oil Emulsions: What's the Difference?*
*Listen to the audio clip and fill in the blanks with the missing information.*

4.2　Section 3

There are two types of emulsion: oil-in-water (O/W) creams and water-in-oil (W/O) creams. Technically, when generating oil-in-water or water-in-oil emulsions, one phase, known as the 1) _____, is mixed into the other, the 2) _____. So, oil-in-water emulsions are composed of small droplets of oil dispersed in a continuous water phase, and in water-in-oil emulsions, the roles of oil and water switch. The result of an emulsion of oil and water mix is dependent on the 3) _____ of both phases and the 4) _____ utilized. The difference can be told by measuring the 5) _____ of the emulsion because water-in-oil emulsions are not conductive.

## Oil-in-Water Emulsions

Oil-in-water emulsions are used in moisturizing products and food products such as milk and mayonnaise. Containing a 6) _____, they are mixable with water, non-greasy, non-occlusive and will absorb water. Water is the dispersion medium in these emulsions and oil-in-water emulsifiers keep oil drops packed in water.

## Water-in-Oil Emulsions

These emulsions are utilized in products like cold cream and are especially useful in products designed for 7) _____. Water-in-oil emulsions are also referred to as 8) _____ sometimes. They mix more easily with oils and have a high oil concentration. Lecithin, lanolin and sorbitan stearate are common water-in-oil emulsifiers.

## Stable Emulsions Are Key

Numerous industries rely on the ability to create stable emulsions efficiently. Without stability, the two phases will separate, and the product will have lower function. Although it's possible to make a temporary emulsion by energetically mixing liquids, the phases will quickly separate if droplet sizes are not 9) _____. Therefore, to achieve stability, emulsifiers are needed. Oil-in-water emulsions typically require more than one emulsifier, and they can be acquired separately or in a mixture. Water-in-oil emulsions only require one emulsifier, but the choices are limited because the 10) _____ must be in a narrow range.

> **Notes**
>
> Most emulsifiers can be considered surfactants or surface-active agents, which can reduce the surface tension of water. What makes an emulsifier surface active is related to its HLB, or hydrophile-lipophile balance. Skin-care emulsifiers can be divided into two groups based on ionic charge: ionic and nonionic. Nonionic emulsifiers such as Glyceryl esters and PEG esters are often used in skincare emulsions for their safety and low reactivity.

### Section 4　The Best Lotion Ingredients

*Listen to the audio clip and complete the following sentences.*

4.2　Section 4

1. Essential fatty acids _____.
   A) can be naturally produced by human bodies
   B) help the cells undergo certain biological process
   C) are the best ingredients for people with acne-prone skin

2. Glycerin _____.
   A) is a kind of occlusive
   B) is composed of sphingosine and a fatty acid
   C) can absorb water from the air

3. Hyaluronic acid _____.
   A) helps collagen and elastin function
   B) can be obtained from food like olive oil and avocado
   C) is a common ingredient in hydrating soaps

4. According to the speaker, people who have eczema and psoriasis _____.
   A) should avoid using products that contain HA
   B) have fewer ceramides than people with normal skin do
   C) should supplement their diet with food rich in fatty acids

5. Statement _____ can NOT be learnt from what you heard.
   A) Ceramides are a family of lipid molecules.
   B) Hyaluronic acid has excellent water-retaining property.
   C) People with oily skin do not need supplement of fatty acids.

**Notes**

Ceramides, a family of lipid molecules, are translated as "神经酰胺" in Chinese, with the suffix "-amide" meaning "酰胺", as in another word niacinamide (烟酰胺). Chemical compounds that belong to the same family usually have the same word roots, such as "-ane" for "烷", "-ene" for "烯", "-one" for "酮", "-ate" for "酸盐", "-yne" for "炔" and "benz" for "苯".

4.2   Further Listening

**Further Listening**

***Why Are Anti-Aging Creams So Expensive?***
*Listen to the audio clip and complete the following sentences.*

## Words and Expressions

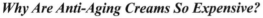

| | | |
|---|---|---|
| rejuvenating *adj.* 焕肤的 | placebo *n.* 安慰剂 | budge item 平价品 |
| glycerin *n.* 甘油 | extravagant *adj.* 奢华的 | broth *n.* 精粹 |
| sea kelp *n.* 海藻 | demonize *v.* 妖魔化 | face-lift *n.* 拉皮 |
| Botox *n.* 肉毒杆菌 | laser surgery 激光手术 | injectable *n.* 注射物 |
| clinical population 临床人群 | smoke and mirror 障眼法 | sea kelp *n.* 海带 |
| unleash *v.* 释放 | | |

1. "Miracle Broth" _____.
   A) is a trademarked ingredient of Cream de la Mer
   B) is extracted from giant sea animals
   C) claims to be able to whiten the skin

2. Cosmetic companies _____.
   A) are obliged to prove the anti-aging efficacy of certain ingredients
   B) must have the FDA approval for the safety and effectiveness of their products
   C) often promote their products by conducting clinical tests which are not credible

3. The audio says that over half of the price that customers pay possibly goes to _____.
   A) extravagant ingredients
   B) the research and development of formulas
   C) the fancy packaging

4. Women of the French court used aged wine as _____.

    A) acid

    B) exfoliant

    C) moisturizer

5. From the audio we can learn that _____.

    A) many anti-aging products are just placeboes

    B) anti-aging products are more popular than cosmetic procedures among the young

    C) a good quality sunscreen is equally important to the expensive ingredients

## Speaking

### Section 1    Eye Cream

*Read the paragraph below and interpret it into Chinese.*

An eye cream is a specially formulated moisturizer that in most cases has been tested as effective to use near the eyes. It won't damage the soft tissue around the eyes or cause eye irritation. Many of these creams are made with special ingredients that help reduce the look of wrinkles around the eyes, provide anti-aging benefits, or help to reduce darker skin tone around the eyes. Eye cream formulas often have heavier moisturizing ingredients, and some night creams can be used safely around the eyes. It is advised to apply any of these creams carefully since even if they don't damage the eyes, they may still cause irritation if some of the cream gets into the eyes. Eye creams may be sold as oils or serums instead of creams, which some people find easier to apply.

### Section 2    Think and Discuss

*Work in a group. Discuss the following questions and share your answers.*

1. What creams or lotions do you use? What are their main ingredients?
2. Among various moisturizers, some lotions or creams are taken as the classic of all time and sold better than others. What are they? What are the reasons for their popularity?

## Critical Thinking

### Project 1    Is Eye Cream a Myth?

Many people believe that eye creams are specially formulated for use around the delicate eye area, while others argue that there is no evidence, research, or documentation validating the claim that the eye area needs ingredients different from those used on the face. Conduct a research on this controversy and present your findings.

*Project 2    Can Hand Cream Be Used on Your Face?*

Suppose you are not at home and you suddenly notice that your face feels dry and irritated. You desperately need some moisturizing lotion, but all you have in your handbag is a tube of hand lotion. Can you use hand cream as face moisturizer? Validate your views with evidence.

## Glossary

1. benzyl alcohol /ˈbenzil ˈælkəˌhɔl/ 苯甲醇
2. butylene glycol /ˈbjuːtiliːn ˈɡlaɪˌkɔl / 丁二醇
3. ceramide /ˈserəmaɪd/ *n.* 神经酰胺
4. continuous phase 连续相
5. cream /kriːm/ *n.* 乳霜
6. dispersed phase 分散相
7. electrical conductivity 导电性
8. emulsifier /ɪˈmʌlsɪfaɪə(r)/ *n.* 乳化剂
9. ethanol /ˈeθənɒl/ *n.* 乙醇
10. fatty acid /ˌfæti ˈæsɪd/ 脂肪酸
11. glycol /ˈɡlaɪˌkɔl/ *n.* 乙二醇
12. HA (hyaluronic acid) 透明质酸
13. hydrophilic /ˌhaɪdrəˈfɪlɪk/ *adj.* 亲水的
14. invert emulsion 反向乳状液
15. lanolin /ˈlænəlɪn/ *n.* 羊毛脂
16. lecithin /ˈlesɪθɪn/ *n.* 卵磷脂
17. lotion /ˈləʊʃn/ *n.* 乳液
18. methanol /ˈmeθənɒl/ *n.* 甲醇
19. O/W emulsion 乳状液
20. polyol /ˈpɒlɪɒl/ *n.* 多元醇
21. propylene glycol /ˈprəʊpiliːn ˈɡlaɪˌkɒl/ 丙二醇
22. sorbitan stearate 山梨坦硬脂酸酯
23. sphingosine /ˈsfɪŋɡəʊsɪ(ː)n/ *n.* 鞘氨醇
24. volume fraction 体积分数
25. W/O emulsion 乳状液

4.2    Keys and Scripts

# Unit 5 Skin Cleansing Products

## Lesson 1 | Body and Facial Cleansing Products

### Checklist for Students

**Key Concepts**

| | | | | |
|---|---|---|---|---|
| bubble bath | shower gel | shower cream | surfactant | synthetic soap |
| thickener | soap bar | foam stabilizer | abrasive | preservative |
| wipe-off cleanser | | water-soluble cleanser | skin conditioning agent | |

**Learning Objectives**

- List the basic types of body cleansing products.
- Understand the major ingredients of body cleansing products.
- Learn what makes good facial cleansers.
- Explain how to choose facial cleansers according to different skin types.

### Before Listening

Cleansing products can be categorized in several ways, including their cleansing mechanism, chemical nature, harshness (mildness), and dosage form. In general, skin cleansers are used to remove dirt, makeup, environmental pollutants, germs, and other types of soilage from the skin. Skin cleansing products contain surfactants that are capable of emulsifying water-insoluble ingredients into micelles, which can be easily washed away from the skin. Ideally, cleansers should not damage the skin's complex structure or lead to irritation, dryness, redness, or itching. Unfortunately, many skin cleansers do cause changes in the skin's structure and barrier function, causing various undesirable signs and symptoms.

### Listening

*Section 1   Types of Body Cleansing Products*
*Listen to the audio clip and fill in the blanks with the missing information.*

5.1   Section 1

Cleansing products for the body are classified into products used in the

1) _____ and products used in the 2) _____. Body cleansers are usually in the form of 3) _____ such as bubble bath, shower gel and shower cream. Such cleansers have 4) _____, because they often contain moisturizing ingredients such as emollients. Compared with simple soap bars, liquid body washes can gradually enhance the skin texture in using, as they clean the skin with 5) _____ _____ and deposit 6) _____on the skin.

Soap bars, on the other hand, were traditionally used to clean the body. They usually contain the ingredients of natural soap, such as 7) _____. However, the product has lost its popularity among people because of some obvious disadvantages. When used in bathtubs with the presence of hard water, natural surfactants will form 8) _____ _____ that can hardly be washed off. Another disadvantage is that unsightly stains of soap will leave on the clothes, which stiffen the fabric.

Compared with soap bars, liquid cleansing products are more convenient and hygienic, providing a 9) _____. Among the several liquid cleansing products, bubble bath is the most popular one. As the name suggests, it can 10) _____ _____. Shower gels are welcomed because they are 11) _____. These transparent gel-like products often include inert particles for 12) _____ _____. Shower creams look like milk or cream and therefore have an 13) _____ _____ appearance. They are usually 14) _____ _____. However, shower gels and creams are different from bubble bath in that they do not produce as much foam, and they have a 15) _____, too.

---

**Notes**

Alkali, an example of natural soap, gets its name from Arabic: al-qaly, meaning "the calcinated ashes". The word alkali or the adjective alkaline is frequently used to refer to all bases. Alkalies are a group of chemicals of high pH. The original production source of alkaline substances was ashes used together with animal fat to produce soap, a process known as saponification.

---

*Section 2   Ingredients Used in Bath Products*
*Listen to the audio clip and make brief summaries of the main ingredients.*

📖 **Overview**

*The types of ingredients used in bubble bath products, shower gels, and creams are generally the same, which include surfactants, thickeners, foam stabilizers, skin conditioning agents, abrasives, and preservatives, etc.*

5.1   Section 2

## Surfactants

*Anionic Surfactants: Features & Functions*

*Nonionic and Amphoteric Surfactants: Features & Functions*

## Thickeners

*Features & Functions*

## Foam Stabilizers

*Features & Functions*

## Skin Conditioning Agents

*Features & Functions*

## Abrasives

*Features & Functions*

## Preservatives

*Features & Functions*

> **Notes**
>
> Surfactants are chemicals that lower the surface tension of a liquid to allow easier spreading. Each surfactant molecule has a hydrophilic component and a hydrophobic component. Surfactants work in three ways: roll up, emulsification and solubilization.

5.1   Section 3

### Section 3   Choosing the Right Facial Cleansers
*Listen to the audio clip and tell if the following statements are true (T) or false (F).*

1. Cleaning face with a wipe-off or cold-cream-type cleanser will leave a film on the skin and prevent it from being damaged by environmental pollutants.
2. Cleansers with super powerful cleaning effect can produce adverse effects such as irritation and barrier damage.
3. A mild water-soluble cleanser is only suitable for dry skin.
4. A qualified water-soluble cleanser should be hydrating, refreshing, and non-irritating.
5. Abrasives and scrubs can be used in facial cleaning twice a day for exfoliating.

5.1   Section 4

### Section 4   Cleansers for Different Skin Types
*Listen to the audio clip and complete the following chart.*

📖 **Overview**
*A right cleanser should clean the skin and leave a soft, pleasant and non-irritated feel. It*

*is especially important to choose facial cleansers according to different skin types, because each skin type reacts differently to the various ingredients in the products.*

| Skin Types | Cleansing Products | |
| --- | --- | --- |
| | Ingredients | Functions |
| normal | retinol, salicylic acid, vitamin C, and chemical exfoliants | speed up skin cell renewal; improve hydration and keep the skin healthy |
| oily | | |
| dry | | |
| combination & sensitive | | |

5.1 Further Listening

## Further Listening

### What If You Stopped Washing Your Face for a Month?

*Listen to the audio clip and give short answers to the questions below.*

## Words and Expressions

remnant *n.* 残余部分                    microbe *n.* 微生物
paradoxically *adv.* 自相矛盾地          micellar *adj.* 胶束的

1. Why is washing face important even if you don't wear any makeup?

_____

2. What factors affect the consequences of not washing face?

_____

3. According to the speaker, what is the correct way of applying "waterless washing"?

_____

4. What benefit of waterless washing has been proved by the research from the University of Michigan?

_____

5. What cleanser options do you have if you want to wash without water?

_____

6. What suggestions are given to make waterless washing work for different individuals?

## Speaking

### Section 1    Micellar Cleansing Water
*Read the paragraph below and interpret it into Chinese.*

When enough surfactant is added to water (more than something called the critical micelle concentration or CMC), the surfactant molecules assemble themselves into clusters called micelles. These micelles are spherical arrangements of surfactant molecules, with the tails pointing in and the heads facing out, which means the hydrophobic tails are protected from the water by the hydrophilic heads. Some brands of micellar water contain oily substances. In these products, the oily substance will sit in the middle of the micelle, like in the emulsion droplet. The micelles are not bound together into a molecule, which means they'll rearrange easily. If the micellar water is poured onto a cotton wool pad, for example, it rearranges so that the heads are stuck to the cotton wool, and the tails stick out. This means there's a neat oil-loving layer sitting on the cotton wool pad. Since make-up, sebum, grease and dirt are all oily, this means micelle water is perfect for cleaning the face as it dissolves the oily substances.

### Section 2    Think and Discuss
*Work in a group. Discuss the following questions and share your answers.*

1. What are your skin cleaning habits and facial cleanser choices?
2. Which do you prefer, body wash or soap? Why?

## Critical Thinking

### Project 1    Soap-based Cleanser or Amino Acid Cleanser?

Normally, facial cleanser focuses on the cleansing efficiency, such as reducing excessive oil, but most of the products irritate the skin. Nowadays, there is an increasing demand for milder cleansers with gentler formulations. What are the differences between soap-based and amino acid cleansers? Which one would you choose, and what are the reasons for your choice?

### Project 2    Safeguard Fined for Claiming "Removing 99% Bacteria"

Safeguard, a famous brand for cleansing products, recently received a fine of 200,000 yuan for using description of "removing 99% of bacteria" on the package of its signature soap product, which, according to the authorities, is an act of false advertising. What are your opinions on it?

## Glossary

1. abrasive /əˈbreɪsɪv/ *n.* 磨料
2. alkali salt 碱金属盐
3. amphoteric surfactant 两性表面活性剂
4. anionic surfactant 阴离子表面活性剂
5. body wash 沐浴露
6. bubble bath 泡沫浴液
7. chemical exfoliant 化学去角质成分
8. cocamide DEA 椰子油二乙醇酰胺
9. fat-soluble /fæt ˈsɒljəbl/ *adj.* 脂溶性的
10. fatty acid 脂肪酸
11. foam stabilizer 泡沫稳定剂
12. inert particle 惰性粒子
13. jojoba oil *n.* 霍霍巴油
14. micelle /maɪˈsel/ *n.* 胶束
15. nonionic surfactant 非离子表面活性剂

16. preservative /prɪˈzɜːvətɪv/ *n.* 防腐剂
17. shower gel 沐浴啫喱
18. skin conditioning ingredients 皮肤调理剂
19. surfactant /sɜːˈfæktənt/ *n.* 表面活性剂
20. tea tree oil 茶树油
21. thickener /ˈθɪkənə(r)/ *n.* 增稠剂
22. water-soluble cleanser 水溶性洁面乳
23. wipe-off cleanser 擦拭型洁面产品

5.1 Keys and Scripts

## Lesson 2 | Makeup Removing Products

### Checklist for Students

#### Key Concepts

| | | | |
|---|---|---|---|
| makeup remover water | makeup remover oil | makeup remover wipe | oil-based remover |
| water-based remover | alcohol-based remover | eye makeup remover | harsh ingredient |
| organic solvent | polyol | emulsifier | mineral oil |
| vegetable oil | synthetic ester | | |

#### Learning Objectives

● List the main types of makeup removers available today.
● Learn the major ingredients used in makeup removing products.
● Differentiate between oil-based remover and water-based remover.

● Discuss the pros and cons of makeup remover wipes.

● Explain how to choose the right makeup remover.

## Before Listening

It is tempting to skip removing makeup after a long day's work or a night party. But leaving makeup on the face overnight can cause the lashes to dry out, clog the pores and lead to acne breakouts. Therefore, it is advisable to remove all the makeup for a refreshed skin the next day. Normally, the removal routine starts with removing eye makeup, then foundation and blush, and finally lip color. There are also a range of makeup removing products for people to choose from, which can be broadly classified into three forms: water, cream and oil. Each product has its own advantages and disadvantages, and therefore is suitable for different skin types.

## Listening

### Section 1    Types and Ingredients of Makeup Removers
*Listen to the audio clip and fill in the blanks with the missing information.*

5.2   Section 1

The importance of makeup removal can never be overestimated, as the failure to remove thoroughly can cause hair keratosis, acne and so on. It is crucial to choose the right kind of products according to the ingredients that 1) _____ _____. The mainstream makeup removing products on the market include 2) _____ _____, 3) _____ and 4) _____ designed with a makeup remover solution. The main ingredients for makeup removers usually include: 5) _____, such as water, which is used in water-based formulations; 6) _____ such as mineral oil; 7) _____ such as poloxamer; 8) _____ _____ such as glycerin; 9) _____ _____ such as carbomers; 10) _____ such as disodium phosphate; 11) _____ such as the parabens; 12) _____ _____ such as calcium disodium EDTA and citric acid; and 13) _____ _____, such as aloe extract and cucumber extract. Makeup removers are made based on the solution formulation or emulsification process. Firstly, 14) _____ are dissolved in water, during which some surfactants and fragrances are added. In the final step, 15) _____ is added when emulsification is finished after careful stirring and mixing.

**Notes**

A carbomer is a homopolymer of acrylic acid, which is cross-linked, or bonded, with any of several polyalcohol allyl ethers. Usually appearing as a white powder, the compound is used as a thickener and emulsion stabilizer. It is consistent, hypoallergenic, and will not support bacterial growth. It also has a particularly nice "skin feel", producing solutions and gels that feel rich and luxurious to the touch. As an emulsion stabilizer, carbomers keep oils or creams suspended in water and prevent separation.

### Section 2　Makeup Remover: Water VS Oil

*Listen to the audio clip and complete the following chart.*

5.2　Section 2

📖 **Overview**

　　*Oil-based and water-based removers are two popular makeup removing products in the market. But special attention is needed in choosing the right type of product for particular needs. Some removers can be too irritating for more delicate areas, such as the eyes and lips. Some others do not have an effective cleaning power to remove the makeup thoroughly.*

| Differences | Oil-based | Water-based |
|---|---|---|
| ingredients | | |
| skin type | | |
| advantages | | |
| disadvantages | | not mentioned |

### Section 3　Makeup Remover Wipes: Pros and Cons

*Listen to the audio clip and give short answers to the questions below.*

5.2　Section 3

1. Why are makeup remover wipes developed?

_____

_____

2. What is the definition of makeup remover wipes?

_____

_____

3. What is the major advantage of makeup remover wipes?

_____

_____

4. Why can't makeup remover wipes be used on a daily basis?

_____

_____

5. Why is it necessary to wash the skin with water after using makeup remover wipes?

_____

_____

6. What are one-time makeup remover wipes made of ?

_____

_____

**Notes**

Makeup remover wipes are more popular in Europe and the United States than in Asia. Perhaps that is because the cuticle of Europeans and Americans is thicker than that of Asians, and most of them tend to have oily skin, for which makeup remover wipes are more suitable and comfortable. Another tip to note is that the special packaging of makeup remover wipes requires added preservatives to extend their shelf life, so the skin can also be irritated by chemicals in them, such as formaldehyde.

*Section 4    How to Pick Up the Right Makeup Remover*
*Listen to the audio clip and tell if the following statements are true (T) or false (F).*

5.2    Section 4

1. Makeup removers with salicylic acid or benzoyl peroxide are suitable for oily or combination skin because they can clean excess oil.
2. Foam-free makeup remover is the best choice for dry skin.
3. Makeup remover oils can help clean up the pores.
4. Remover with an emulsifying creamy formula is suitable for sensitive skin.
5. Alcohol is an ingredient to avoid in makeup removing products, because it cannot break down the tough chemical components in makeup.
6. Baby oil is recommended as makeup remover because it is super mild.
7. Ingredients such as fragrances and preservatives might lead to irritation.

5.2 Further Listening

## Further Listening

### *The Importance of Removing Makeup*
*Listen to the audio clip and complete the following chart.*

### Words and Expressions

regenerate *v.* 再生                      flaccidity *n.* 松弛                      premature *adj.* 提早的

antioxidant *n.* 抗氧化剂                 allergic *adj.* 过敏的                    eczema *n.* 湿疹

pigmentation *n.* 色素沉着               chronic *adj.* 慢性的，长期的            dehydrate *v.* 使脱水

chapped *adj.* 有裂痕的，皲裂的          puffy *adj.* 肿胀的                       micellar water 胶束水

| Importance | Details |
| --- | --- |
| Delay aging process | Scientist have proven that your skin can 1) _____ itself. If it's not clean, it will 2) _____. |
| Nourish face | You can help the skin cells to reproduce by removing makeup and applying 3) _____. Using 4) _____ is also vital to have healthy skin. |
| Fight allergies | Sleeping with makeup can cause 5) _____ and other health problems with symptoms like 6) _____. |
| Avoid skin pigmentation | The dark spots on the face can be the consequences of 7) _____. |
| Keep lips hydrated | 8) _____ can continue to dehydrate the delicate skin and lead to dry lips. |
| Prevent bags under eyes | If you often leave your makeup on, you will soon have 9) _____ under your eyes. |
| Prevent eyelashes from falling out | To remove mascara, you'd better use 10) _____ but not 11) _____. |
| Prevent acne | If you suffer or have suffered from acne, it's good for you to use 12) _____ to clean your face. |

Speaking

### Section 1   Application Tips of Eye Makeup Removers
*Read the paragraph below and interpret it into Chinese.*

Eye makeup removers are very necessary for cleaning makeup in the delicate eye area, but they have to be used carefully. Avoid pulling the skin repeatedly, as it stretches the skin and damages the elastin fibers. Elastin, which gives healthy skin the flexible and resilient feature, is a form of stretchable protein found in skin tissues. The pulling in cleaning eye makeup with removers will change its orderly arrangement, no matter to which direction the skin is pulled: up, down or to the side. Even if your makeup removing procedures are very gentle, pulling is unavoidable and will inevitably stretch the skin. When you see the skin move, you know the elastic fibers are being tugged, which quickens the sagging of skin. It is advised to watch closely into the mirror when removing eye makeup to avoid pulling as much as possible to prevent wrinkling and damaging the skin.

### Section 2   Think and Discuss
*Work in a group. Discuss the following questions and share your answers.*

1. What are your makeup removing procedures? Are there any useful tips?
2. Recommend makeup removers to your partner based on his or her skin type.

Critical Thinking

### Project 1   Is It Necessary to Use Different Removers for Different Areas?

Makeup removers can be classified according to the areas of skin they are applied to, such as facial makeup remover and makeup remover for the eyes and lips. Are there any differences between these makeup removers in terms of functions and ingredients? Is it necessary to develop removers specifically for different areas of skin?

### Project 2   How to Remove the Makeup in Traditional Chinese Operas

It is recorded in many ancient Chinese medical books and literary works that the Chinese people had rich experiences in removing facial makeup with various natural ingredients, such as rice washing water, soaps made from herbal formulas, and vegetable oils for heavier makeups, such as makeup in traditional Chinese operas. How are the removers in ancient China different from modern removers in terms of ingredients and formulations? What is the best way to remove heavy makeup?

## Glossary

1. almond oil 杏仁油
2. aloe extract 芦荟提取物
3. calcium disodium EDTA 二钠钙
4. carbomer /ˈkɑːbəmə(r)/ *n.* 卡波姆
5. chelating agent 络合剂，螯合剂
6. citric acid 柠檬酸
7. color additive 色素
8. cucumber extract 黄瓜提取物
9. disodium phosphate 磷酸二钠
10. elastin /ɪˈlæstɪn/ *n.* 弹性蛋白
11. emulsification process 乳化过程
12. isopropyl alcohol 异丙醇
13. keratosis /ˌkerəˈtəʊsɪs/ *n.* 角化病
14. non-woven cloth 无纺布
15. oil-based remover 卸妆油

16. organic solvent 有机溶剂
17. pH buffer pH 缓冲剂
18. poloxamer /pɒˈlɒksəmə/ *n.* 帕洛沙姆
19. polyester /ˌpɒliˈestə(r)/ *n.* 聚酯纤维
20. solubilizer /ˈsɒljʊbɪlaɪzə(r)/ *n.* 增溶剂
21. solution formulation 溶剂配制
22. water-based remover 卸妆水
23. waterproof products 防水产品

5.2 Keys and Scripts

# Unit 6  Hair Care Products

## Lesson 1  Hair Cleansing Products

### Checklist for Students

#### Key Concepts

| | | | |
|---|---|---|---|
| emulsification | liquid | gel emulsion | thickener |
| anti-dandruff shampoo | cleansing agent | foaming agent | chelating agent |
| vitamin B group | skin irritation | eye irritation | zinc pyrithione |
| ketoconazole | selenium sulfide | sulfur derivative | surfactant-based preparation |

#### Learning Objectives

- Describe briefly the history of using shampoo.
- List the main types of ingredients used in shampoos.
- Learn the major functions of shampoos.
- Understand how shampoos can negatively affect the scalp and hair.
- Name some active ingredients that can be used in anti-dandruff shampoos.

### Before Listening

Shampoo is a basic hair care product representing the largest segment of hair care cosmetics. Shampoo is typically in the form of a viscous liquid with some exception of waterless solid form such as a bar. It was developed to replace soap for cleansing scalp and hair. Shampoo typically consists of an aqueous emulsion of several surfactants (at least one of which has strong detergent properties), various additives to enhance cleansing performance or improve sensory attributes, one or more preservatives, and fragrances. Anti-dandruff shampoos can contain a number of active ingredients, which function through very different mechanisms.

## Listening

### Section 1　The History of Using Shampoos
*Listen to the audio clip and choose the correct answer to each of the following questions.*

6.1　Section 1

1. What do ancient Egyptians use to wash their hair?
   A) Grape seeds and kohl.
   B) Citrus juice and soap.
   C) Olive extracts and tea tree oil.

2. What is the origin of "shampoo"?
   A) It originates from India and refers to a hair-washing massage with special ingredients like alkali, natural oils and fragrances.
   B) It originates from British salons and refers to a beautifying hair styling product.
   C) It originates from America after WWII and refers to a soap-based hair wash made with aromatic herbs.

3. What is the significance of the invention of soapless detergent?
   A) It improves the cleaning power of shampoos.
   B) It reduces the allergy caused by shampoos.
   C) It cleans the hair without leaving residue.

4. What are the major formulas of shampoos on the market nowadays?
   A) Formulas for dry, damaged and oily hair.
   B) Formulas for normal, colored and permed hair.
   C) All of the above.

5. What additional benefits do some special modern shampoos claim to offer?
   A) Volumizing and straightening.
   B) Volumizing and scalp massaging.
   C) Styling and preventing gray hair.

### Section 2　Ingredients of Shampoo
*Listen to the audio clip and fill in the blanks with the missing information.*

6.1　Section 2

　　Shampoo products on the present market come in many forms, such as 1) _____ _____. But in terms of chemical nature, shampoos are basically 2) _____. The major ingredients include 3) _____, with other components like foaming agents, chelating agents, pearlescents, dyes, fragrances, and preservatives.

　　Cleansing agents alone will cause hair to become dull, coarse and hard to comb and manage, which is why other additives are included in the formulations to 4) _____. One example of such special substances are 5) _____ _____ and others. If some ingredients are in fashion, companies will cater to consumer needs and add them in the formula to increase sales volume.

　　Among the many additives, the vitamin B group, especially vitamin B5 and B6, are at the center of interest. According to some cosmetic and pharmaceutical companies, if used on a regular basis,

products containing these vitamin substances can 6) _____.
Another benefit is that hair will gradually become supple and strong. But this claim has not been validated by scientific research or studies, and the effects yet need to be further tested. Since the hair shaft is made of 7) _____, the external part of hair can neither be 8) _____, or grow to be healthier simply by 9) _____. But shampoos with vitamin substances do 10) _____ _____. In this aspect, so long as the ingredients help to improve the appearance and manageability of hair, consumers with particular hair care needs will invariably prefer such preparations.

**Notes**

Pearling agents are added to shampoos to increase the sensory acceptability and to impart the perception of richness. To obtain a satisfactory pearl effect in shampoo formulas, concentrated blends of ethoxylated and non-ethoxylated alkyl stearates are generally used.

### Section 3　Functions and Adverse Reactions of Shampoos
*Listen to the audio clip and tell if the following statements are true (T) or false (F).*

6.1　Section 3

1. The main function of shampooing is to remove dirt from hair, which is made up of sweat, sebum and its breakdown products, dead skin cells, residues of cosmetics and personal care products, dust and other environmental impurities in the air.
2. Surfactants in shampoos can only remove water-soluble substances.
3. Shampoos could irritate the skin when applied to the scalp and hair while massaging and rinsing.
4. Eye irritation from shampoos is caused by the primary surfactants, such as sodium lauryl sulfate.
5. Ingredients such as amphoteric surfactants, silicone derivatives, and protein derivatives are added in shampoos to reduce eye irritation.
6. It is advised to wash the hair frequently to remove sebum as much as possible for both oily hair and dry hair.

### Section 4　Dandruff and Anti-dandruff Shampoo
*Listen to the audio clip and give short answers to the questions below.*

6.1　Section 4

1. What is the definition of dandruff?
_____
_____

2. Why does dandruff appear?
_____
_____

3. What are the common symptoms of dandruff?
_____
_____

4. What are the two types of dandruff and what are their respective characteristics?

_____

_____

5. What factors could contribute to the formation of dandruff?

_____

_____

6. What are the functions of anti-dandruff shampoos?

_____

_____

7. What are the tips in using anti-dandruff shampoos?

_____

_____

_____

### Notes

Malassezia refers to a group of club-shaped yeasts of the genus Malassezia, which has several different species. As lipophilic fungi, they colonize skin rich in sebaceous glands. The skin fungus Malassezia lives on everyone's scalp — both people with and without dandruff. In some people, the Malassezia fungus breaks down the scalp skin barrier, making an oleic acid by-product as a result. This acid can irritate the skin and trigger an inflammatory response, causing scalp cells to clump and flake off.

## Further Listening

### Dry Shampoo

*Listen to the audio clip and give short answers to the questions below.*

6.1  Further Listening

## Words and Expressions

| | |
|---|---|
| ingenious *adj.* 独创的，巧妙的 | talc *n.* 滑石粉 |
| cetrimonium chloride 西曲氯铵 | tress *n.* 一缕头发 |

1. When do people need to use dry shampoo?

_____

2. What are the main ingredients of dry shampoo?

_____

3. What problems can be caused by consistent use of dry shampoo?

_____

4. What harmful ingredients are there in dry shampoo?

_____

5. Which is recommended, dry shampoo or the traditional way of washing hair? Why?

_____

## Speaking

### Section 1   Fragrance and Shampoo

*Read the paragraph below and interpret it into Chinese.*

The use of fragrance has a dramatic effect on the commercial success of shampoo. It has been said that it is the advertising, package, fragrance and color, rather than the quality and performance of the product that sell the first bottle of shampoo. When selecting fragrance, it is of top importance that the fragrance-developing company has access to the shampoo base as well as to the fragrance profile. Fragrances can affect shampoo viscosity, color stability, odor stability and clarity. All of these factors need to be considered when developing, selecting and testing a fragrance. Fragrances are also a major cost contributor to a shampoo and therefore they are typically used at low concentrations of 0.2% to 1.0%. In many cases, antioxidants and UV absorbers are used in shampoos to protect the color stability.

### Section 2   Think and Discuss

*Work in a group. Discuss the following questions and share your answers.*

1. What is the best type of shampoo for you?
2. What do you think of hair loss prevention shampoo?

## Critical Thinking

### Project 1   Shampoos: the More Expensive, the Better?

Can you believe that a bottle of shampoo of 200 ml is sold for more than 100 RMB? The rumor that P&G will introduce its high-end shampoo is spreading in the industry. Meanwhile, domestic company Bawang also announced to release a Chinese Shampoo Package for 168 RMB. Is it true that the more expensive the shampoo is, the better functions it has? What factors contribute to the rise of a high-end shampoo market?

***Project 2    Silicone-free Shampoo***

It is said that if you are prone to hair loss, oil, dandruff or dry hair, it's because you have been using silicone-containing shampoo, and it's time to change to a silicone-free shampoo. How can you tell if a shampoo is silicone-free? Is it better than silicone-containing shampoo? Make an analysis.

## Glossary

1. aqueous emulsion 水乳液
2. aromatic herbs 芳香植物
3. cell turnover 细胞更新
4. coal tar *n.* 煤焦油
5. dandruff /'dændrʌf/ *n.* 头皮屑
6. detergent /dɪ'tɜːdʒənt/ *n.* 清洁剂
7. dye /daɪ/ *n.* 染料
8. flaking /'fleɪkɪŋ/ *adj.* 起屑的
9. foaming agents *n.* 发泡剂
10. keratin /'kerətɪn/ *n.* 角质蛋白
11. ketoconazole /ˌkitəu'kəunəzəul/ *n.* 酮康唑
12. Malassezia 马拉色菌
13. odor /'əudə(r)/ *n.* 气味
14. pearlescent /pə'lesnt/ *n.* 珠光成分
15. residue /'rezɪdjuː/ *n.* 残留
16. scale /skeɪl/ *n.* 剥落
17. selenium sulfide *n.* 硫化硒

18. sensory attributes 感官特征
19. silicone derivatives 硅衍生物
20. soap base 皂基
21. sodium lauryl sulfate 十二烷基硫酸钠
22. static electricity *n.* 静电
23. sulfur /'sʌlfə(r)/ *n.* 硫
24. viscosity /vɪ'skɒsəti/ *n.* 稠度
25. vitamin B group B 族维生素
26. volumizing /'vɒljʊmaɪzɪŋ/ *adj.* 使头发丰盈的
27. zinc pyrithione *n.* 吡啶硫酮锌

6.1  Keys and Scripts

## Lesson 2 | Hair Conditioning Products

### Checklist for Students

**Key Concepts**

| | | |
|---|---|---|
| hair damage | properties of hair | cationic detergent |
| quaternary conditioner | film-forming conditioner | instant conditioner |
| deep conditioner | protein-containing conditioner | hair rinse |
| static electricity | polyvinylpyrrolidone (PVP) | leave-in product |

**Learning Objectives**

- Understand the definition, action and functions of hair conditioners.
- Tell the differences of major types of hair conditioners available today.
- Learn the ingredients commonly used in hair conditioners.
- Discuss proper ways of using hair conditioners.
- Explain the roles of silicones in hair conditioners.

### Before Listening

Hair conditioner is a hair care product that is applied after shampooing in order to condition the hair. It is most useful for people with dry or damaged hair. People with naturally oily hair may find that conditioners weigh their hair down rather than improve the overall look and feel of it. There are a wide range of hair conditioning products, including those you rinse out, leave in, or spray on. Available in a wide range of prices, hair conditioner is essential to healthy hair, though various hair types require different treatments. Dry hair or hair damaged from chemical treatments, sun, or heat styling should be conditioned with each shampooing. Fine, untreated, and oily hair can also benefit from hair conditioner, but with less frequent applications.

### Listening

**Section 1   *Hair Conditioners: Definition, Action and Functions***
*Listen to the audio clip and fill in the blanks with the missing information.*

Conditioners are applied to the hair after shampooing and are designed to smooth the hair, 1) _____, as well as recondition chemically

6.2   Section 1

damaged hair by 2) _____, mechanically damaged hair by 3) _____, and weathered hair by 4) _____. Conditioners act by 5) _____ generated after combing dry hair, improving manageability by 6) _____ around and between the cuticle scales, increasing hair shine by 7) _____ with a thin layer, decreasing 8) _____, and improving hair flexibility. The coat that they create covers the outer, rough layer of the hair, giving the hair its smooth, uniform look by 9) _____ on the surface of the hair. By doing this, the hair does not look so unruly, and becomes much easier to comb and style; it also 10) _____. 11) _____ is the ideal conditioner. Excessive removal of sebum leads to a harsh and dull appearance of the hair. Therefore, it is necessary to use 12) _____ sebum-like products.

### Section 2    Types of Hair Conditioners

*Listen to the audio clip and complete the following chart.*

6.2    Section 2

| Conditioners | Formulations | Ingredients | Application | Hair type |
|---|---|---|---|---|
| Instant conditioners | lotion | quats | leave for several minutes before rinsing off | frequent washing or little hair damage |
| Hair rinses | | quats, such as stearalkonium chloride | | |
| Deep conditioners | | quats and hydrolyzed proteins | | |
| Leave-in conditioners | | mineral oil and silicones | | |

**Notes**

The most common type of cationic surfactant used in conditioners are quaternary ammonium ions, nicknamed "quats". Cationic surfactants comprise a long chain hydrocarbon as the lipophile, a quaternary amine nitrogen as hydrophile, and a halide ion as counterion. Cationics are attracted to surfaces carrying a negative charge, upon which they adsorb strongly. Proteins and synthetic polymers can thus be treated with cationics to provide desirable surface characteristics.

6.2   Section 3

### *Section 3   Ingredients of Hair Conditioners*
*Listen to the audio clip. Match the descriptions with different types of conditioners.*

📖 **Overview**

   Hair conditioners on the market come in liquids, creams and gels. They mainly contain conditioning agents, such as lipids, silicones, protein derivatives, and glycols. The most commonly used ingredients are quaternary conditioners, film-forming conditioners, protein-containing conditioners and silicones.

A) quaternary conditioners
B) film-forming conditioners
C) protein-containing conditioners
D) silicone-containing conditioners

1. _____ They are mainly PVP to coat the hair fibers with a thin layer of polymer so as to fill in the defects in the cuticle.
2. _____ They contain a small amount of protein fragments broken from animal tissues, silk and plants to penetrate hair shaft and improve its strength.
3. _____ They are detergents with cationic compounds to neutralize the negative charges in the hair.
4. _____ They work well for straightening curly hair, but they weigh down naturally straight hair and make it difficult to style.
5. _____ They form a thin film on the hair without producing a greasy and limp look.
6. _____ They work well on damaged hair, such as colored and permed hair.

---

**Notes**

PVP, also called povidone or polyvinylpyrrolidone, is used in the formulation of a wide range of cosmetic products including mascara, eyeliner, hair conditioners, hair sprays, and shampoos. PVP helps to distribute or to suspend an insoluble solid in a liquid and keeps emulsions from separating into their oil and liquid components. It dries to form a thin coating on the skin, hair or nails.

---

6.2   Section 4

### *Section 4   How to Use Hair Conditioners*
*Listen to the audio clip and tell if the following statements are true (T) or false (F).*

1. It is not advised to use hair conditioner every day.
2. Too much and too little conditioning are both undesirable practices in hair care.
3. Conditioners should always be used after shampooing.
4. Co-wash refers to hair treatment that combines several conditioning products in hair care.
5. Washing your hair while conditioning it can address problems such as dryness, fragility and friction.

6. One can sleep with conditioner left on the hair overnight.

7. Continuous expanding and shrinking of hair cuticle lead to hygral weariness.

8. The cuticle becomes stronger after a large amount of water absorption.

6.2  Further Listening

## Further Listening

### *Silicone in Hair Conditioners*
*Listen to the audio clip and give short answers to the questions below.*

### Words and Expressions

sulfate *n.* 硫酸盐　　　　plague *n.* 瘟疫，灾祸　　　lubricant *n.* 润滑剂

sealant *n.* 密封剂　　　　insulation *n.* 隔热，绝缘　　dimethicone *n.* 二甲聚硅氧烷

veneer *n.* 虚饰　　　　　limp *adj.* 柔软的　　　　　lackluster *adj.* 无光泽的

castor *n.* 蓖麻　　　　　argan oil 摩洛哥坚果油　　　cyclomethicone *n.* 环甲基硅酮

1. What is silicone?

_____

2. What products can silicones be used in? Write at least three of them.

_____

3. What are the common silicone compounds found in hair conditioners?

_____

4. Why are silicone-containing hair products great for styling?

_____

5. According to the manufacturers of the hair product "Noughty", what would happen if silicone builds up on the hair?

_____

6. What are the functions of sulfates in shampoos with silicones?

_____

7. What does the speaker say about water soluble silicones?

_____

8. What are the natural alternatives of silicones? Write at least three of them.

_____

## Speaking

### Section 1　*This Is What Happens When You Stop Using Conditioners*
*Read the paragraph below and interpret it into Chinese.*

One problem of hair conditioner is that for the purpose of moisturization, it usually coats the hair strands, which can make the hair look soft and limp. When you stop conditioning, the hair feels lighter and the scalp is free from conditioner residues. For people with such concerns, it does not harm to skip hair conditioning once in a while. In the first few weeks, the hair may look smooth and shiny, but over time, without proper conditioning and managing, the hair will become a tangled mess. Adam Friedman, a Washington, D.C.-based board-certified dermatologist says in an interview that "conditioner refortifies the cuticle with a protective coating, allowing the hair to keep growing and not break easily. When the hair is exposed to the outside world, the cuticle, or outer lining, gets damaged until it ultimately breaks; the conditioner fills in those injuries and coats the hair to assist the cuticle." Hair conditioning is particularly helpful for dry and damaged hair, because it can smooth up the unruly or parched strands. Therefore, for easier combing, more lustrous look and healthier hair, hair conditioning should never be avoided completely.

### Section 2　*Think and Discuss*
*Work in a group. Discuss the following questions and share your answers.*

1. Do you have the habit of hair conditioning? Do you think it is useful?
2. What hair conditioners do you use? What are the differences between them?

## Critical Thinking

### Project 1　*Hair Conditioner or Hair Oil?*

In the competition for the best hair care, hair conditioners are always highly rated among endless hair products. But some people question whether conditioners can provide as comprehensive effects as a well-matched hair oil does. What are the strengths and weaknesses of each product? How to choose the best hair conditioning product for different individuals?

### Project 2　*Keratin Hair Treatment*

A keratin hair treatment can be either a salon treatment or an at-home procedure that uses proteins to smooth the hair cuticle and create straighter, shinier hair. The protein keratin is naturally found in human hair, and keratin treatments are aimed at restoring this protein to frizzy, curly or damaged hair. How is keratin hair treatment done? How effective is it? Are there any side effects?

## Glossary

1. cationic /ˌkætˈaɪənɪk/ *n.* 阳离子
2. coating /ˈkəʊtɪŋ/ *n.* 涂层
3. co-wash /ˈkəʊwɒʃ/ *n.* 护发素洗头法
4. cuticle scale 角质层鳞屑
5. electric charge *n.* 电荷，电极
6. film /fɪlm/ *n.* 膜
7. hair mask 发膜
8. hair rinse 冲洗型护发素
9. hydrolyzed proteins 水解蛋白质
10. instant conditioner 即用型护发素
11. leave-in hair conditioner 免洗护发素
12. lipid /ˈlɪpɪd/ *n.* 脂质，脂类
13. negative charge *n.* 负（阴）电荷
14. neutralize /ˈnjuːtrəlaɪz/ *v.* 中和
15. polymer /ˈpɒlɪmə(r)/ *n.* 聚合物

16. PVP (polyvinylpyrrolidone) 聚乙烯吡咯烷酮
17. quats (quaternary ammonium compound) 季铵化合物
18. residue /ˈrezɪdjuː/ *n.* 残留物
19. silicone /ˈsɪlɪkəʊn/ *n.* 硅
20. split end 分叉
21. static electricity *n.* 静电
22. stearalkonium chloride 十八烷基二甲基苄基氯化铵

6.2 Keys and Scripts

# Unit 7　Makeup Products

## Checklist for Students

### Key Concepts

| | | | |
|---|---|---|---|
| lip gloss | lip balm | lip stick | lip liner |
| matte | glaze | shimmer | fragrance |
| preservative | fixative | wax | oil |
| antioxidant | coloring agent | texturizing agent | fat and butter |
| pigment premilling | melting and mixing | molding and packaging | flaming |
| aeration laddering | chipping | cratering | streaking |
| sweating | bleeding | | |

### Learning Objectives

- Differentiate between lipstick, lip gloss, lip liner and lip balm.
- List the typical types of lip makeup products and provide some examples.
- Describe the technique "molding" and explain its four basic steps.
- Understand some of the typical quality issues that may occur during the formulation of lip makeup products and explain why they may occur.

## Before Listening

Lip makeup products mainly include lip sticks, lip glosses, lip liners and lip balms. They are applied to enhance or color the lips and are purely for increasing beauty and attractiveness. Lipsticks and lip balms are produced with a technique called molding, which includes four steps: pigment premilling, melting and mixing, molding and packaging, and flaming. The typical quality-related issues of lip care products include aeration, laddering, chipping, deformation, cratering, streaking, sweating and bleeding.

## Listening

### Section 1    Categorization of Lip Makeup Products
*Listen to the audio clip and match the product types with the definitions.*

7.1    Section 1

1. Lipstick _____
2. Lip gloss _____
3. Lip liner _____
4. Matte lipstick _____
5. Glossy or glaze lipstick _____
6. Crème lipstick _____
7. Shimmer lipstick _____
8. Moisturizing lipstick _____

A) It is a type of cosmetic product that adds color to the lips while providing moisture at the same time. It includes softening and moisturizing ingredients such as vitamin E, oils such as macadamia oil, or shea butter.

B) It is a type of cosmetic application for lips which is known for its ability to last all day and its lack of shine. It is rich in pigments and waxes, but contains less oil, which results in the dark finish.

C) It is a small stick of waxy lip coloring enclosed in a cylindrical case. It is applied to enhance the appearance of the lips by giving it color and gloss.

D) It is a cosmetic product used to outline and define the lips. Available from dozens of brands and in a rainbow of colors, it can be applied to create a subtle or bold statement. It mostly resembles a pen or pencil in design.

E) It is designed to give the lips a glossy luster and, sometimes, subtle color. It usually has a lower viscosity than traditional lipsticks and is more transparent.

F) It usually falls between mattes and glosses. It contains a high concentration of emollients to create a shiny, glossy finish.

G) It is heavy in oil and often contains flavors and scents as well. It adds shine and volume to the lips but wears off quickly.

H) It contains light-reflecting particles, such as coated micas to add luster to the color. It is more often used for special occasions and comes in lighter colors.

### Section 2    Typical Ingredients of Lip Makeup Products
*Listen to the audio clip and fill in the blanks with the missing information.*

7.1    Section 2

    The major ingredients used in lip makeup products can be classified into several basic categories. Waxes function as 1) _____, providing lipsticks with 2) _____. When combined with different properties, waxes can help to achieve 3) _____, such as high shine, flexibility, and brittleness. Oils, fats, and butters provide a 4) _____ _____ to the formulations. They also have a 5) _____ and act as emollients. Oils are also used to 6) _____. Butters and fatty acid esters can improve the 7) _____ for the formulations to the lips. Color additives distinguish lipsticks from other lip products, giving the products commercial and appearance value. Main color additives used in colored lip cosmetics include 8) _____ _____, lakes, and specific or the so-called 9) _____ _____. The function of antioxidants is to prevent 10) _____ of sensitive ingredients. Preservatives are added to prevent the products from being contaminated by 11) _____. Fragrances are used to 12) _____ _____, therefore, they must have an agreeable taste without causing irritation. Texturizing ingredients, such as talc, silica, and mica, may also be used to improve the 13) _____ _____ of products. In some cases, fixatives are added to 14) _____ _____ on the lips. Lip makeup products may also contain active ingredients, such as 15) _____ _____.

## Notes

Lake pigments (lakes) are a type of organic pigment made by precipitating a dye with an inert binder, or "mordant", which is usually a metallic salt. Lakes are not oil soluble but are oil dispersible. As comparatively inert absorption compounds, they are highly versatile and adaptable. Lake colors are typically used to make coated tablets, cake and doughnut mixes, hard candies and chewing gums, lipsticks, soaps, shampoos, talc, etc.

### Section 3　Formulation Techniques of Lipsticks
*Listen to the audio clip and make brief summaries of formulation techniques.*

#### 📖 Overview
　　*There are various techniques available for the formulation of lip makeup products. The most frequently applied technique to produce lipsticks and lip balms is the process known as molding. The usual steps of this technique include pigment premilling, melting and mixing, molding and packaging, and flaming.*

7.1　Section 3

**Step 1. Pigment premilling**

**Step 2. Melting and mixing**

**Step 3. Molding and packaging**

**Step 4. Flaming**

> **Notes**
>
> During lipstick formulation, it is important to minimize air incorporation into the mass. Pinholes will appear in the stick during molding if air is not removed from the bulk, slowing down production and increasing rejection rate. It can be eliminated by slow and uninterrupted filling. Overmixing should also be avoided.

7.1　Section 4

*Section 4　Quality-related Issues of Lip Makeup Products*
*Listen to the audio clip and complete the following chart.*

| Problems | Definition or Causes |
|---|---|
| Aeration | |
| Laddering | |
| Chipping | Chipping or cracking may occur if the lipstick mix is too brittle. They may be caused by an imbalanced wax/oil ratio or a faulty cooling technique. |
| Cratering | |
| Streaking | Streaking refers to the appearance of discontinuities and streaks, that is, a thin line or band of different color substances on the stick's surface. |
| Sweating | |
| Bleeding | |

## Further Listening

7.1　Further Listening

*The Evolution of Red Lips*
*Listen to the audio clip and complete the following sentences.*

### Words and Expressions

defiant *adj.* 反叛的　　　　status *n.* 地位　　　　reputation *n.* 名声
incarnation *n.* 化身　　　　reign *n.* 统治　　　　toxic *adj.* 有毒的
deem *v.* 认为　　　　　　　patriotic *adj.* 爱国的　　progressive *adj.* 进步的
transgressive *adj.* 跨越的　myriad *adj.* 无数的

1. During the Dark Ages, wearers of red lipsticks were regarded as _____.
　 A) people with higher economic status

B) representations of the devil

C) esteemed aristocrats

2. Queen Elizabeth I liked red lipsticks because she believed that _____.

A) they had protecting and healing effect

B) they can ward off evil spirits

C) they can add beauty to the face

3. In the 1900s, people wore red lipsticks in marches as a sign of _____.

A) independence            B) equality            C) rebellion

4. Red lips have influenced the post-war Hollywood in the way that _____.

A) They became the signature of stars and punctuated many pop culture moments.

B) They became household necessities for urban women.

C) They became tools to create artistic works.

## Speaking

### Section 1    Choosing the Best Lipstick Color

*Read the paragraph below and interpret it into Chinese.*

The lipstick shade that you pick should compliment your skin tone, as well as incorporate with your hair color, eye color and natural lip color. All your efforts of applying lipstick perfectly are of no use if the shade of your lipstick is mistaken. The right shade makes a lot of difference in the appearance and overall personality you want to present. If you have dark skin you have the greatest flexibility: plums, chocolates, reds and oranges all work with your skin. Generally, the deeper your skin tone, the deeper the shade of lipstick you can wear. If you have a medium skin tone with golden undertones, all shades of red are for you. And lastly if you have pale skin, stick to nudes, beige tones, light corals and light pinks.

### Section 2    Think and Discuss

*Role-play as customer and shop assistant at a lipstick counter and make a conversation that includes the following topics.*

1. types, brands and features of various lipstick products
2. application tips and makeup tutorials of lipsticks, etc.

## Critical Thinking

### Project 1    Reviving Traditional Chinese Lip Makeup Culture

If eyes are the window to the soul, lips are the mirrors of one's character and temperament. Being an important part of face decoration in ancient China, lip makeup enjoys a long history and has various patterns in different periods. What are the features and patterns of lip makeup in different dynasties of ancient China? How can the traditional makeup culture and heritage be rejuvenated in the modern context?

*Project 2    Safety Concerns of Lip Makeup Products*

A study involving 22 brands of lipsticks found that 55% contained small amounts of lead which could lead to women's mental problem. Some researchers claimed that even low exposure causes risks that could affect human health in multiple ways. Are the harmful effects of chemicals in lipsticks real? How to improve the safety standards of lip makeup products?

## Glossary

1. aeration /eə'reɪʃ(ə)n/ *n.* 口红膏体出现小气泡
2. bleeding /'bliːdɪŋ/ *n.* 口红晕色
3. chipping /'tʃɪpɪŋ/ *n.* 口红膏体有划痕
4. cracking /'krækɪŋ/ *n.* 口红膏体断裂
5. cratering /'kreɪtərɪŋ/ *n.* 口红膏体表面有坑洞
6. crème lipstick 柔润唇膏
7. disperse /dɪ'spɜːs/ *v.* 分散
8. effect pigment 效应颜料
9. glaze lipstick 唇釉
10. homogeneous /ˌhəʊmə'dʒiːniəs/ *adj.* 均质的
11. inorganic pigment 无机颜料
12. laddering /'lædərɪŋ/ *n.* 口红膏体分层
13. lakes /leɪks/ *n.* 色淀（颜料）
14. lip gloss *n.* 唇彩
15. lip liner *n.* 唇线笔
16. lipstick /'lɪpstɪk/ *n.* 口红
17. matte /mæt/ *n.* 哑光
18. melted phase 熔融相
19. mica /'maɪkə/ *n.* 云母
20. moisturizing lipstick 保湿唇膏
21. molding /'məʊldɪŋ/ *n.* 灌装成模
22. nude /nuːd/ *adj.* 裸色的
23. organic color 有机颜料
24. pestle and mortar 研磨钵
25. pigment dispersion 颜料分散液
26. pinholing /'pɪnhəʊlɪŋ/ *n.* 口红膏体有小孔
27. premilling /'priːmɪlɪŋ/ *n.* 预磨
28. shimmer lipstick 珠光唇膏
29. silica /'sɪlɪkə/ *n.* 硅石，二氧化硅
30. streaking /'striːkɪŋ/ *n.* 口红膏体有条纹痕迹
31. structuring agent 结构支撑剂
32. sweating /'swetɪŋ/ *n.* 口红膏体结油珠
33. talc /tælk/ *n.* 滑石粉
34. texturizing ingredients 质地形成剂
35. UV filter UV 滤光剂
36. wax /wæks/ *n.* 蜡

7.1  Keys and Scripts

## Lesson 2　Facial Makeup Products

### Checklist for Students

**Key Concepts**

| | | | |
|---|---|---|---|
| foundation | concealer | BB cream | blush |
| compact powder | oil-based | water-based | oil-free |
| water-free | antioxidant | emulsification | preservative |
| thickening agent | chelating agent | direct pigment | pigment blend |
| monochromatic color solution | | liquid foundation | powder foundation |
| cream-to-powder foundation | | | |

**Learning Objectives**

- List the typical types of facial makeup products and give some examples.
- Differentiate between foundation, blush and concealer.
- List the main ingredient types in powder facial makeup products and understand cosmetic formulations in powder form.
- List the main ingredient types in liquid and semi-solid facial makeup products and explain their formulation techniques.
- Learn the common mistakes in applying foundation and the correct way of blending foundation.

### Before Listening

Facial makeup products are intended to conceal small blemishes on the facial skin and give it a more refined, better textured, yet still natural look. Major facial makeup products on the market include foundations, concealers and blushes, which can be listed as cosmetics or OTC drug-cosmetic products in the US. The products can come in powder forms, such as loose powders and compact powders, or liquid and semisolid formulations, such as lotions and creams. Additional product types available on the market today include sticks, two-way foundations, cream-to-powder and cream-in-powder foundations, transparent facial powders, and primers.

## Listening

### Section 1   Types and Definitions of Facial Makeup Products
*Listen to the audio clip and fill in the blanks with the missing information.*

7.2   Section 1

Facial foundations are designed to create a 1) _____ _____, provide a 2) _____ to the skin, as well as 3) _____ for all skin tones. They are usually applied to the whole facial area. Foundations come in various types and forms to suit different consumer needs. Foundations are available as 4) _____, which represent the most popular form of facial makeup products, as well as 5) _____.

Concealers, also known as 6) _____, are very similar to foundations in terms of their ingredients. However, they are designed to hide minor skin imperfections, such as 7) _____. Concealers are more noticeable than foundations, as they have much 8) _____ pigments. Another difference is that concealers are primarily applied to areas of concern on the face, rather than the whole facial skin. They are available in various forms, including 9) _____.

Blush, or rouge, is designed to 10) _____. However, to many customers, blush denotes a 11) _____ product, which is more popular, while rouge denotes a 12) _____ product. They are usually applied 13) _____ to the cheek area of the face to emphasize and highlight the cheekbones by adding more color. Powder blush is similar to facial powder in formulation, except with 14) _____. Cream rouges are similar to 15) _____, which contain light esters, waxes, mineral oil, titanium dioxide, and pigments.

### Section 2   Powder Facial Makeup Products and Ingredients
*Listen to the audio clip and complete the following chart.*

7.2   Section 2

📖 **Overview**

*Facial powder makeup products, including facial foundation, concealer, and blush, are available as loose powders and compact powders. These products are identical to powder eyeshadows in terms of the ingredient types and formulation technology. The main ingredients are as follows:*

| Ingredients | Examples | Functions |
|---|---|---|
| Fillers | talc | |
| Absorbents | kaolin<br>starch<br>chalk | |

(continued)

| Ingredients | Examples | Functions |
|---|---|---|
| Binders | zinc<br>starches<br>oils | |
| Colorants | iron oxides<br>ultramarine<br>chrome hydrate<br>chrome oxide<br>pearls | |
| Preservatives | phenoxyethanol | |
| Antioxidants | BHA, BHT, vitamin E | |

**Notes**

BHA (butylated hydroxyanisole) and BHT (butylated hydroxytoluene) are used as preservatives in many foods, cosmetic products and drugs. In cosmetics, these ingredients are found mainly in shampoos, deodorants, body lotions and make-up, usually at a concentration of 0.1% or less. BHA and BHT play an important part in maintaining the quality and safety of products, and help to extend shelf life.

*Section 3  Liquid and Semi-solid Facial Makeup Products and Formulations*
*Listen to the audio clip and briefly summarize the main features of each formulation.*

7.2  Section 3

📖 **Overview**

*Liquid and semisolid makeup formulations are primarily emulsions in the form of lotions, creams, and mousses. They can be oil-based, water-based, oil-free, or water-free formulations.*

**Oil-based formulations (example)**

**Type**: W/O emulsions
**Ingredients**: suspended pigments, emollients, occlusive, water, emulsifier, vitamin, sunscreens, moisturizing agents
**Features**: They do not shift color, but may feel greasy and heavy if not used by one with dry skin.

## Water-based formulations

Type:

Ingredients:

Features:

## Oil-free formulations

Type:

Ingredients:

Features:

## Water-free formulations

Type:

Ingredients:

Features:

### Notes

As liquid and semisolid facial makeup products are emulsions, their manufacturing is an emulsification process. The formulation procedures mainly include pigment premilling, dispersion with oil and mixing into the oil phase. The emulsion base is colorated in different ways, and the shade of foundations often has to be corrected.

7.2 Section 4

### Section 4 *How to Blend Foundation*

*Listen to the audio clip and tell if the following statements are true (T) or false (F).*

1. A thin application of foundation with sunscreen effect can provide adequate protection from the sun.

2. The best place to check the blending technique is in the bathroom.

3. The areas that you miss in applying foundation, such as the ears, mouth, often look streaked, or appear blotchy or smudged in daylight.

4. It is recommended that the best way to blend foundation is using fingers.

5. For an even application, shake some foundation from the bottle onto the face and then dab with a sponge.

6. In applying foundation, one should hold the sponge in the fingers, spread it with a stroking, buffing motion, going in the direction of the hair growth.

**Notes**

Brushes and sponges should be cleaned at least once a month. All it takes is a gentle shampoo, warm water and a mug. Swirl the brushes around the water and massage the bristles to get out all of that residue that would otherwise be rubbed onto your face. Rinse, squeeze, and lay out to dry. It's a good idea to do this before bed so they're ready to be used again by the morning.

**Further Listening**

7.2 Further Listening

### *Cushion Compacts*

*Listen to the audio clip and complete the following sentences.*

## Words and Expressions

sheer *adj.* 轻薄的    airtight *adj.* 密封的    applicator *n.* 粉扑    sponge *n.* 海绵
fair *adj.* 肤色白皙的    flip *v.* 翻转    skin tone 肤色    matte *adj.* 哑光的
discoloration *n.* 斑点    dullness *n.* 暗沉    neutralize *v.* 中和    primer *n.* 妆前乳
toner *n.* 化妆水    jawline *n.* 下颌线    dewy *adj.* 湿润的    finish *n.* 质感

1. The packaging of cushion compacts is bulkier than that of average ones because _____.

    A) it creates lightweight sheer coverage.

    B) it makes an airtight seal to keep the cushion fresh.

    C) it makes it easier to carry around.

2. The key to choosing the right cushion compacts lies in _____.

    A) selecting the right finish and coverage

    B) flipping the compact over before using

    C) choosing the correct applicator

3. The color correction shade of purple can _____.

    A) help with coloration

    B) combat dullness

    C) brighten dark circles

4. Cushion compacts are not the best choice for oily skins because _____.

    A) they cannot offer even coverage.

    B) they are not water resistant.

    C) they have dewy finishes.

5. _____ is the tip for choosing the right cushion compacts for acne-prone and mature skin types.

    A) Look for added skincare benefits and avoid heavy formula with harsh ingredients.

    B) Use hydrating setting mist before the application.

    C) Apply heavier layers of the product to cover fine lines and skin blemishes.

## Speaking

### Section 1    Foundation Makeup Tips

*Read the paragraph below and interpret it into Chinese.*

The method of applying foundation has certainly progressed in recent years. It also depends on the type of look you're going for. For example, if you prefer a natural look, you may want to know how to apply powder foundation rather than liquid. The tools and techniques differ depending on your choice of coverage, texture and whether or not you're sticking to your true skin tone. Liquid foundations can be the easiest to leave a few streaks. To avoid these streaks, we can use foundation brush and sponge. If you prefer your natural complexion to shine through but still want to blur imperfections and even out your skin tone, you may need a powder foundation. To ensure that your powder foundation doesn't fall into any cracks, it's best to apply a primer first to create a smooth base to work with. You also want to make sure it doesn't end up looking cakey or dry, which is largely due to lack of hydration.

### Section 2    Think and Discuss

*Discuss how to choose the right foundation from the following aspects.*

1. Features and functions of various foundation products.

2. Skin types and skincare concerns of individuals, etc.

## Critical Thinking

### *Project 1 Ancient Roman Makeup VS Tang Dynasty Makeup*

Both ancient Rome and Chang'an celebrate long history of facial makeup traditions, each influenced by unique civilizations at that time. However, the styles of two makeups are quite different, especially in the use of foundation and blush. Explore the characteristics of ancient Roman and Chinese facial makeups and make a comparison.

### *Project 2 Market Trend of Foundation Products*

As the key to a perfect makeup, foundations have always been the most important type among facial makeup products. With upgrading consumer needs and changing makeup aesthetics, foundations are met with new expectations and challenges from the market, such as skincare benefits, longer wearing time, etc. Analyze the current market needs and predict the future development trend of foundation products.

## Glossary

1. absorbent /əbˈzɔːbənt/ *n.* 吸附剂
2. anhydrous /ænˈhaɪdrəs/ *adj.* 无水的
3. BHA 丁基羟基茴香醚
4. BHT 二丁基羟基甲苯
5. binder /ˈbaɪndə(r)/ *n.* 黏合剂
6. blemish /ˈblemɪʃ/ *n.* 瑕疵，斑点
7. blush /blʌʃ/ *n.* 腮红
8. chalk /tʃɔːk/ *n.* 白垩土
9. chrome oxide 氧化铬
10. colorant /ˈkʌlərənt/ *n.* 着色剂
11. concealer /kənˈsiːlə(r)/ *n.* 遮瑕
12. coverage /ˈkʌvərɪdʒ/ *n.* 遮盖力
13. cream-in-powder foundation 粉凝霜粉底
14. cream-to-powder foundation 霜转粉粉底
15. ester /ˈestə(r)/ *n.* 酯
16. filler /ˈfɪlə(r)/ *n.* 填充剂
17. foundation /faʊnˈdeɪʃn/ *n.* 粉底

18. hydrocarbon /ˌhaɪdrəˈkɑːbən/ *n.* 碳氢化合物，烃类，碳氢
19. iron oxide 氧化铁
20. kaolin /ˈkeɪəlɪn/ *n.* 高岭土，瓷土
21. liquid foundation 粉底液
22. loose powder 散粉
23. mousse foundation 慕斯粉底
24. opaque /əʊˈpeɪk/ *adj.* 不透明的
25. primer /ˈpraɪmə(r)/ *n.* 妆前乳
26. silicone-based 硅基的
27. sponge /spʌndʒ/ *n.* 上妆海绵
28. starch /stɑːtʃ/ *n.* 淀粉
29. stick foundation 棒状粉底
30. suspended pigment 悬浮颜料
31. synthetic ester 合成酯
32. titanium dioxide 二氧化钛
33. transparent facial powder 透明蜜粉

34. two-way foundation 干湿两用粉底
35. ultramarine /ˌʌltrəməˈriːn/ *n.* 群青色
36. water-in-silicone emulsion 硅油包水乳状液
37. zinc /zɪŋk/ *n.* 锌

7.2　Keys and Scripts

## Lesson 3　Eye Makeup Products

### Checklist for Students

**Key Concepts**

| | | | | |
|---|---|---|---|---|
| mascara | eyeliner | eyebrow liner | eyeshadow | eyeshadow brushe |
| filler | absorbent | binder | colorant | color extension |
| blending | milling | filling | sieving | compression |
| cream-to-powder eyeshadow | | gel eyeshadow | | cream eyeshadow |

**Learning Objectives**

● List the major types of eye makeup products and explain their respective features and functions.
● Learn the basic ingredients of mascara and the formulation techniques.
● Learn the basic ingredients of eyeshadow and the formulation techniques.
● Learn the correct application techniques of eyeshadow.

### Before Listening

Makeup for the eye has been controversial in history, as the eyes are regarded as symbols of mystery and lust in different cultures and times. In modern times, eye makeup was one of the last face cosmetic products to become popularized and mass marketed. Eye makeup products include those that are used around the eye to enhance its appearance and emphasize the beauty. The main products are eyeshadow, eye liner, eyebrow products, and others to enhance and accent the eyes.

## Listening

7.3　Section 1

### Section 1　Types and Definition of Eye Makeup Products

*Listen to the audio clip and complete the following chart.*

| Types | Functions | Compositions | Forms |
|---|---|---|---|
| Mascara | produce an intense look; make the eyelashes thicker, longer, and darker; highlights and dramatizes the eyes | waxes, pigments, texturizers, emulsifiers, solvents | liquid form cake form |
| Eyeliner | | | |
| Eyebrow liner | | | |
| Eyeshadow | | | |
| Eye makeup remover | | | |

### Notes

Many people have eye infections from contact and use of eye makeup products. At the time of purchase, most eye makeup products are free from bacteria. But if they are not properly preserved, or misused by consumers after being opened, problems may happen, as dangerous bacteria are allowed to enter and grow in the product. Discontinue immediately the use of any eye product that causes irritation. It should also be recognized that bacteria on the hands could, if placed in the eye, cause infections. Frequent handwashing is recommended before applying eye makeup.

### Section 2　Mascara: Ingredients and Formulations

*Listen to the audio clip and fill in the blanks with the missing information.*

7.3　Section 2

#### 📖 Overview

　　*Cosmetics have been used to darken the eyelashes and eyebrows in many cultures since ancient times. Eugene Rimmel, a French perfume magnate, was the first to commercially market mascara in the 1830s. The next significant industry innovation took place in 1913, when T. L. Williams, the founder of Maybelline cosmetics, combined coal dust and petroleum jelly to produce cake mascara that was applied to the lashes with a moistened brush. Helena Rubenstein pioneered the now ubiquitous tube-*

*and-wand version in 1957. Today, two main types of mascara are marketed: cake mascara and the more popular liquid mascara. Mascaras can also be classified as water-resistant and waterproof types.*

Cake mascara has a 1) _____ texture and is applied to the eyelashes via 2) _____. It smudges easily 3) _____, because most formulations have little water 4) _____. The soap/wax/pigment blend is emulsified when a wet brush is applied to its surface. The main ingredients for cake mascara include 5) _____ _____.

Water-resistant liquid mascara is a typical 6) _____, while waterproof liquid mascara is typically 7) _____ formulations made via dispersing waxes in non-aqueous solvents, which provide water-proofing effect and contribute to the 8) _____. Ingredients for the two products are mainly similar, which include solvents, structurants, 9) _____, emulsifiers, color additives, 10) _____, preservatives, antioxidants, chelating agents, emollients, etc.

Water-based mascara is generally oil in water emulsion. In the formulation process, the water-soluble thickeners are 11) _____ in water and 12) _____ if necessary. After the hydration process, all water-soluble components are added to the 13) _____. Afterwards, waxes, oils and other emollients are mixed and heated until the ingredients melt. When both phases 14) _____, they will be blended in small amounts with continuous mixing. In the next step, color additive dispersions are added to the emulsion. When 15) _____, other ingredients such as preservative are added and mixed.

**Notes**

For centuries, a wide-eyed look is always desired. This look has long been achieved by the use of mascara and additional cosmetics. Some people turn to eyelash perming, or keratin lash lift, for lashes that point upwards. The perm is performed by placing a perming solution on the eyelashes and wrapping them around a roller. It is important to note, however, that eyelash perming does not make eyelashes darker, thicker, or longer. It can be dangerous to the delicate eye area, too.

### Section 3　*Eyeshadow: Ingredients and Formulations*
*Listen to the audio clip and give short answers to the questions below.*

7.3　Section 3

**Overview**
*Eyeshadows are available as pressed and loose powders, creams, sticks, and pencils. Loose and pressed powder eyeshadows are the most popular forms of this product category. Aside from powder eyeshadows, which are by far the most common type, other types available include liquids, pencils, cream-to-powder, and creams. Though these can be fun and easy to use, they are often hard to blend and control.*

1. What are the main ingredients for both pressed and loose powders?

2. What is the difference between the manufacturing steps of loose and pressed eyeshadows?

3. Why are milling and blending necessary in eyeshadow formulation?

4. Why are anhydrous eyeshadows often called "cream-to-powder" eyeshadows?

5. What are the basic steps of the formulation of cream eyeshadows?

### Section 4   Eyeshadow Application Techniques
*Listen to the audio clip and choose the correct answer to each of the following questions.*

7.3   Section 4

1. Why is it not advised to use a sponge-tip applicator?
   A) Because they irritate the skin.
   B) Because they tend to blend colors in streaks.
   C) Because they leave a matte finish.
2. What is the correct way to achieve an even and well-blended eyeshadow design?
   A) Laying strips of color that overlap and blend together over the eye.
   B) Beating the brush back and forth across the eye.
   C) Tapping the brush over spots for intense effect.
3. What kind of brush should you use for a larger eyelid area?
   A) Use a small brush with the width of the eyelid.
   B) Use a brush with hard bristles.
   C) Use a wide and full brush.

## Further Listening

### Korean Eye Makeup VS American Eye Makeup
*Listen to the audio clip, draw pictures of the makeups and make brief summaries.*

7.3   Further Listening

## Words and Expressions

defined *adj.* 干净分明的        arched *adj.* 弓形的        concealer *n.* 遮瑕膏

crisp *adj.* 锐利的          lashes *n.* 假睫毛          lash line 眼睑
smoky look 烟熏妆          brow mascara 染眉膏          cat eye look 猫式眼妆

**American Eye Makeup**                    **Korean Eye Makeup**

Eyebrow:                                    Eyebrow:



Eyeshadow:                                  Eyeshadow:



Eyeliner:                                   Eyeliner:



Lashes:                                     Lashes:

## Speaking

### Section 1   Matte Eyeshadow
*Read the paragraph below and interpret it into Chinese.*

Matte powder eyeshadows in an array of neutral tones from light to dark are your best choices for a classic, sophisticated eye design that accents the shape and color of your eyes. Matte refers to the look of the eyeshadow; matte eyeshadow is flat and does not contain any light reflectors, whereas other types of eyeshadow may appear to be shiny. Some people mix different types of eyeshadow for a more customized appearance. Matte eyeshadow comes in many colors, and may be available in a cream or powder form. It is generally worn during the day for a more subtle appearance, whereas shiny or shimmery eyeshadow may be worn more often in the evening to look more glamorous. Both types of eyeshadow typically have the same silky texture to go onto the eyelids smoothly.

### Section 2   Think and Discuss
*Discuss with your partner on how to make the eyes look bigger with various eye makeup products and demonstrate to the class. Your tutorial may include:*

1. Types, features and functions of eye makeup products

2. Application techniques to make the eyes look bigger

3. Discussion of eye makeup styles and aesthetics

## Critical Thinking

*Project 1   Foxy Eyes: New Aesthetics or Racist Stereotypes?*

Models with foxy eyes on fashion magazines are accused of racial discrimination against the Asian people, as slanted eyes that look small, narrow or thin have long been a stereotypical feature of Asian look. How are makeup styles associated with cultural images and how to express the uniqueness and diversity of the beauty of the people through the art of makeup?

*Project 2   Eyelash Extensions and Implants*

Eyelash extensions and implants are gaining more popularity as they can increase the length and volume of natural eye lashes. But it should be noted that such applications may result in certain problems, such as irritation and allergic reactions. How can such makeup products be applied in a safe manner? Probe into this trend and propose your research findings.

## Glossary

1. applicator /ˈæplɪkeɪtə(r)/ *n.* 上妆工具

2. base powder 基底粉体

3. blending /ˈblendɪŋ/ *n.* 混合

4. cake mascara 块状睫毛膏

5. compression /kəmˈpreʃən/ *n.* 压制，压缩

6. eye makeup remover 眼部卸妆产品

7. eyebrow liner 眉笔

8. eyelash /ˈaɪlæʃ/ *n.* 眼睫毛

9. eyeliner /ˈaɪlaɪnə(r)/ *n.* 眼线笔

10. eyeshadow /ˈaɪʃædəʊ/ *n.* 眼影

11. filler /ˈfɪlə(r)/ *n.* 填充剂

12. film-former /film ˈfɔːmə(r)/ *n.* 成膜剂

13. liquid mascara 液体睫毛膏

14. loose eye powder 散粉眼影

15. mascara /mæˈskærə/ *n.* 睫毛膏

16. matte eyeshadow 哑光眼影

17. milling /ˈmɪlɪŋ/ *n.* 研磨

18. pressed eye powder 压制眼影

19. sieving /ˈsɪvɪŋ/ *n.* 筛滤

20. solvent /ˈsɒlvənt/ *n.* 溶剂

21. stabilizer /ˈsteɪbəlaɪzə(r)/ *n.* 稳定剂

22. texturizer /ˈtekstʃəraɪzə(r)/ *n.* 质地形成剂

23. water-proof /ˈwɔːtər pruːf/ *adj.* 防水的

24. water-resistant /ˈwɔːtə rɪzɪstənt/ *adj.* 抗水的

7.3   Keys and Scripts

## Lesson 4   Nail Care Products

### Checklist for Students

**Key Concepts**

| | | | |
|---|---|---|---|
| functional nail product | decorative nail product | technical quality | nail hardener |
| nail moisturizer | nail polish | cuticle remover | color additive |
| thixotropic agent | color stabilizer | pigment preparation | artificial nail |
| base coat | top coat | resin | solvent |
| plasticizer | blending | mixing | bubbling |
| cracking | chipping | thickening | |

**Learning Objectives**

- Differentiate between major types of nail care products.
- Learn the basic ingredients of nail polish and explain their respective functions.
- Understand and describe the formulation of nail polish.
- List and explain the typical quality-related issues of nail care products.

### Before Listening

Nail care products can be classified into functional ones and decorative ones. The former promotes the healthy growth of human nails and helps their normal maintenance and removal of cuticle. Major functional nail products include nail hardeners and cuticle removers. The latter is applied to give nails color, which involves various types of nail polish and artificial nails. Nail care products are also used to treat brittle, soft and splitting nails with nail cosmetics.

### Listening

*Section 1   Basic Types of Nail Products*
*Listen to the audio clip. Match the products with their functions and components.*

1. _____ Nail hardeners
2. _____ Nail moisturizers
3. _____ Nail polishes
4. _____ Cuticle removers
5. _____ Artificial nails
6. _____ Nail polish removers

7.4   Section 1

A) They are organic solvents that are used to remove previously applied nail polish.

B) They aim to improve the hardness of nails with the moisturizing ingredients. They come in product forms of lotions or creams which can be directly applied to the nail plates.

C) They are designed to remove the excess tissue and dead skin cells. They are usually a liquid or cream form with alkali ingredients to destroy the cuticle keratin.

D) They are a type of lacquer used to decorate fingernails and toenails. They consist of pigments suspended in a volatile non-aqueous solvent to which film-formers are added.

E) They are nails created using products designed to enhance nail length. Products in this category include preformed plastics, formed acrylics, and a combination of both.

F) They are designed to provide a protective layer on the nail plate by increasing the hardness and strength of the nails.

---

**Notes**

Cuticle removers should not be confused with cuticle softeners. Softeners are liquids or creams that only wear down the cuticle for easy removal by subsequent cutting or trimming. Cuticle softeners contain quaternary ammonium compounds in a 3%–5% concentration, and they are sometimes combined with urea. For complete disengagement of the excess tissue and dead skin cells, cuticle removers are preferred.

---

### Section 2　Ingredients of Nail Polish

*Listen to the audio clip and fill in the blanks with the missing information.*

7.4　Section 2

　　Nail polish can contain a dozen or more different ingredients. Despite the potentially high number of ingredients and a vast array of colors, ingredients can be categorized into several basic groups.

　　Resins, also called 1) _____, hold the ingredients of the 2) _____ together and form a strong film on the nails. There are two types of resins: hard glossy resins and soft pliable resins. They are used in different concentrations to achieve various effects of the nail polish. Hard glossy resins can form 3) _____, while the soft pliable resins enhance 4) _____. Solvents function as carriers of lacquer. They 5) _____ resins, 6) _____ pigments, and 7) _____, leaving a smooth film. Solvents also prevent the polish from drying out and regulate the 8) _____ of the product. Plasticizers improve resin flexibility and 9) _____. The colors in nail polish come from a pigment called 10) _____. Other ingredients, such as powdered aluminum, mica flakes, and bismuth oxychloride are added to create a shimmer effect. Thixotropic agents, or 11) _____, help disperse the color additives and improve the viscosity of nail polish at rest. Color stabilizers are added to prevent nail polish from having 12) _____ when exposed to UV light. Some top coats contain UV filters to prevent nail polish from 13) _____ over time. Nail treatment ingredients contain components that 14) _____ _____, which may include vitamins, minerals, oils, 15) _____, and fibers, such as silk.

**Notes**

Bismuth oxychloride is a white lustrous powder. It is an inorganic white pigment commonly used in foundations, blushes and other color cosmetics. Bismuth oxychloride usually comes in two forms: diamond and pearl finishes. The diamond is shimmery and the pearl is more matte. Bismuth is the least toxic of its periodic table neighbors like lead, tin, and polonium. Bismuth in itself is not safe for use in cosmetics, and must be refined and combined with other elements to produce bismuth oxychloride.

### Section 3    Formulation of Nail Polish
*Listen to the audio clip and give short answers to the questions below.*

7.4    Section 3

1. Why does the formulation of nail polish products involve a high risk?

_____

2. Why is pigment preparation the most important step in nail polish formulation?

_____

3. What form are nail polish pigments prepared in?

_____

4. How is the polish base made in nail polish formulation?

_____

5. In coloring of the lacquer base, when are other solvents and additional ingredients added?

_____

### Section 4    Quality-related Issues of Nail Care Products
*Listen to the audio clip and complete the following chart.*

7.4    Section 4

| Term | Problem(s) | Cause(s) |
|------|-----------|----------|
| Bubbling | a visible process on the surface of the newly painted nails | air that may come from "overshaked" bottles |
| Cracking | | |

(continued)

| Term | Problem(s) | Cause(s) |
|---|---|---|
| Chipping | | |
| Thickening | | not mentioned |

## Further Listening

7.4 Further Listening

### Nail Polish's Surprising Origins

*Listen to the audio clip and tell if the following statements are true (T) or false (F).*

### Words and Expressions

| | | | |
|---|---|---|---|
| tactic *n.* 战术 | Babylonian *n.* 巴比伦人 | stain *v.* 弄脏 | gum Arabic 阿拉伯树胶 |
| gelatin *n.* 明胶 | orchid *n.* 兰花 | predecessor *n.* 前任 | execute *v.* 处死 |
| salon *n.* 美容院 | buffer *n.* 缓冲剂 | bejeweled *adj.* 装饰珠宝的 | |

1. Around 3200 BC, Babylonian soldiers stained their fingernails to warn each other of danger on the battlefield.
2. The Zhou Dynasty of China preferred a gold and silver nail polish, while the Ming Dynasty preferred red and black.
3. Egyptian queen Cleopatra was obsessed with nail polish and liked to paint her whole hands.
4. Nail polish used to be a priority only for the upper class.
5. In the 19th century, well-manicured hands were a symbol of social status.
6. In 1917, an American company invented the first liquid nail polish from car paint finish.
7. Both dyes and pigments can offer nail polish various shades.

## Speaking

### Section 1　Gel Manicure

*Read the paragraph below and interpret it into Chinese.*

A gel manicure is a type of nail treatment in which a nail technician uses gel to bind synthetic fingernails to natural ones and color them. During a gel manicure, a nail technician first cleans and shapes a person's natural nails and pushes the cuticles back, after which a base coat is applied to help the fake nail or polish adhere better. Several more coats of polish are then applied to the nails. After each layer is painted on, the person getting the manicure puts his or her hands under a small Ultra Violet lamp for a few seconds or minutes to for the nail to cure and harden. The final step is applying a clear top coat to seal and protect the nails. Though gel nails are generally long-lasting and don't chip, people do have to take special care when removing them. Since gel polish is so durable, standard nail polish remover won't work on gel nails and picking them off can damage one's nails. Therefore, harsher chemicals are needed to dissolve and remove the gel.

### Section 2    Think and Discuss

*Suppose you are at a manicure salon to have a nail care. Role-play as customer and manicurist to make a conversation, which may include:*

1. Types, effects and decorations of nail polishes
2. Care tips after manicure

## Critical Thinking

### Project 1    Nail Art in the Qing Dynasty

Fostering nails had been a fad in ancient times, especially for females of wealthy families in ancient China. To protect their slender and fragile fingernails, a sheath was designed particularly for nails. The sheath, in the shape of a bow, was always worn on the ring finger and little finger, and at the top part of a finger. During the Qing Dynasty, the nail sheath was made mostly of gold, silver, jade, hawksbill, pearl and gold-gilded copper, and the craftsmanship was enamel, filigree, carving and diancui (dipping blue). Why do concubines wear fingernail sheaths in the royal palace? What are the functions of such decorations? Conduct a research on this topic and make a presentation.

### Project 2    Safety and Health Concerns of Nail Polish

Is statement "UV filters in nail polish provide nail cancer prevention" a fact or a fiction? Why? Explore other safety and health issues related to nail care products and make a presentation.

## Glossary

1. acetone /ˈæsɪtəʊn/ *n.* 丙酮
2. adhesion /ədˈhiːʒn/ *n.* 黏附力，附着力
3. artificial nails 人造指甲
4. bismuth oxychloride 氯氧化铋

5. bubbling /ˈbʌblɪŋ/ *n.* 指甲油起泡

6. butyl acetate 乙酸丁酯，醋酸丁酯

7. aluminum /əˈluːmɪnəm/ *n.* 铝

8. chipping /ˈtʃɪpɪŋ/ *n.* 指甲油剥落

9. cross-linking agent 交联剂

10. cuticle keratin 角质层蛋白

11. cuticle remover 去指皮水

12. acetate /ˈæsɪteɪt/ *n.* 醋酸盐

13. ethyl acetate 乙酸乙酯，醋酸乙酯

14. evaporate /ɪˈvæpəreɪt/ *v.* 蒸发

15. flammable /ˈflæməbl/ *adj.* 易燃的

16. acrylic /əˈkrɪlɪk/ *n.* 丙烯酸

17. high-shear mixer 高剪切混合器

18. lacquer /ˈlækə(r)/ *n.* 漆

19. lacquer base 漆基，基料

20. mica flake 云母薄片

21. nail hardener 硬甲油，指甲强化剂，
    指甲精华素

22. nail moisturizer 指甲保湿剂

23. nail plate 甲盖

24. nail polish 甲油

25. nail polish base 甲油底胶

26. nail polish remover 洗甲水

27. nitrocellulose /ˌnaɪtrəʊˈsɛljəˌləʊs/ *n.* 硝化
    纤维（素）

28. organic solvent 有机溶剂

29. performed plastics 预制塑料

30. pigment blend 颜料混合物

31. plasticizer /ˈplæstəˌsaɪzə(r)/ *n.* 增塑剂，
    塑化剂

32. pliable /ˈplaɪəbl/ *adj.* 可塑的，易弯曲的

33. resin /ˈrezɪn/ *n.* 树脂

34. resistance performance 阻力性能

35. solvent carrier 溶液载体

36. suspending agent 助悬剂，悬浮剂

37. thixotropic agents 触变剂

38. volatile /ˈvɒlətaɪl/ *adj.* 不稳定的；
    易挥发的

7.4  Keys and
Scripts

# Unit 8　Cosmetic Efficacy

## Lesson 1　Moisturizing

### Checklist for Students

#### Key Concepts

| | | | | | |
|---|---|---|---|---|---|
| NMF | CA | FLG | SC | mineral oil | petrolatum |
| humectant | emollient | occlusive | sorbitol | collagen | antioxidant |
| free radical | retinoid | vitamin C | vitamin E | skin rejuvenator | vegetable fat |
| hyaluronic acid | | alpha-hydroxy acid | | | |

#### Learning Objectives

- Define skin moisturizers and list the major types of them.
- Understand the definition, functions and components of NMF.
- Learn the types, functions and mechanisms of humectants and occlusive.
- Learn common agents added to moisturizers and explain their functions.

### Before Listening

Today, skin moisturizers, protective and nourishing products take the largest market share out of the total sector of skin care and personal care products. Both men and women all over the world have considerable need for moisturizing cosmetics. A study report shows that 75% of the younger generation uses moisturizers on a daily basis. As new ingredients and more advanced technologies are introduced in recent years, moisturizing products now have better functionality and aesthetic appeal, too. Moisturizer products contain ingredients that can keep the skin hydrated, replace the lost NMF in the corneocytes, replenish the skin with intercellular lipids, and form a protective layer over the skin. As the first step in anti-aging skin care, skin moisturization is crucial for healthy skin appearance, softness, and enhancement of the barrier functions against the environmental weathering.

## Listening

### Section 1　Types and Definitions of Skin Moisturizers
*Listen to the audio clip and give short answers to the questions below.*

8.1　Section 1

1. What are the main functions of skin moisturizers?

_____

_____

2. What kind of moisturizing product is considered as drugs in the US? Why?

_____

_____

3. What moisturizing ingredients can be used as OTC for skin protectants?

_____

_____

4. Why is it recommended to use different moisturizers for specific body parts?

_____

_____

5. What are the four major types of skin moisturizers?

_____

_____

### Section 2　Natural Moisturizing Factor (NMF)
*Listen to the audio clip and fill in the blanks with the missing information.*

8.1　Section 2

　　The term "natural moisturizing factor" first appeared in English-language publications in 1959, coined by Jacobi and colleagues. Studies reporting the discovery of the NMF in the epidermis refer to it as "1) _____ " or ingredients whose removal decreased 2) _____. The role of the NMF is to maintain 3) _____, which helps maintain plasticity of the skin, allows hydrolytic enzymes to function in the process of desquamation and contributes to 4) _____. The NMF is composed principally of 5) _____ such as PCA, urocanic acid, and 6) _____, as well as lactic acid and urea. NMF components are highly efficient humectants that attract and bind water from the atmosphere, drawing it into the corneocytes. This process can occur even 7) _____, allowing the corneocytes to maintain an adequate level of water in low-humidity environments. The water absorption is so efficient that the NMF essentially dissolves within the water it has absorbed. Hydrated NMF, particularly the neutral and basic amino acids, forms 8) _____, reducing the intermolecular forces between the fibers and thus increasing the elasticity of the stratum corneum. The NMF may be reduced by decreased production or processing of filaggrin, or by 9) _____. Several NMF components have long been used in moisturizers. Thus, there is evidence to support the incorporation of additional NMF components into treatments for skin conditions that can 10) _____. We can now more fully recognize the importance of the NMF and the beneficial role that these humectant substances can play as cosmetic agents.

---

**Notes**

Filaggrin (FLG) is one of the major skin barrier proteins that form a protective layer at the stratum corneum level of the skin. *FLG* binds keratinocyte filaments to increase the density of filament bundles, flattening keratinocytes to their terminal shape, which is crucial to the strength and integrity of the skin. Filaggrin metabolites work as natural moisturizing factors, which are able to absorb a large amount of water, maintaining the necessary level of humidification of the tissue and the appropriate pH.

8.1   Section 3

*Section 3   Occlusive and Humectants*
*Listen to the audio clip and complete the following chart.*

| Comparison | Occlusive | Humectants |
|---|---|---|
| mechanism | form1) _____<br>provide 2) _____<br>impair 3) _____ | enhance 6) _____<br>absorb 7) _____ |
| effects | particularly effective in the treatment of dry, damaged skin | accelerate maturation of corneocytes, reduce redness, enhance 8) _____ |
| ingredients | petrolatum, 4) _____ _____ paraffin, and squalene | 9) _____, hyaluronic acid and sorbitol. |
| drawbacks | 5) _____ | In relatively dry environment, humectants with large molecules may 10) _____. |

---

**Notes**

Humectants are essential cosmetic ingredients which are hygroscopic in nature and absorb moisture from the air. Humectants form hydrogen bonds with water molecules and help to retain the natural moisture of the skin. Basically, humectants are molecules with one or more hydrophilic groups attached to them. These hydrophilic groups can be either amines, like urea or amino acids, carboxyl groups such as fatty acids and alpha hydroxy acids, or hydroxyl groups such as glycerin, sorbitol and butylene or glycols.

### *Section 4  Additives to Moisturizers*
*Listen to the audio clip and tell if the following statements are true (T) or false (F).*

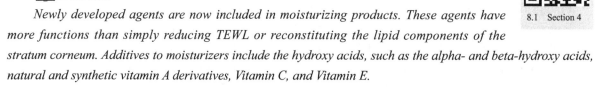

8.1  Section 4

#### 📖 *Overview*
*Newly developed agents are now included in moisturizing products. These agents have more functions than simply reducing TEWL or reconstituting the lipid components of the stratum corneum. Additives to moisturizers include the hydroxy acids, such as the alpha- and beta-hydroxy acids, natural and synthetic vitamin A derivatives, Vitamin C, and Vitamin E.*

1. Alpha-hydroxy acids can promote cell proliferation and increase the moisture level in cell culture.
2. Salicylic acid can increase the peeling-off of the stratum corneum, which makes older skin feel "smoother and fresher".
3. Natural and synthetic vitamin A derivatives, known as retinoids, are often added to moisturizers as water-absorbing agents.
4. Retinoic acid can increase the thickness of the epidermis and promote the deposition of new collagen within the dermis.
5. Vitamin A in high enough concentrations can behave as an occlusive, preventing water from evaporation.
6. Many companies claim that in cell culture studies, vitamin E increases collagen production.
7. Vitamin E is more effective when applied topically than taken orally.
8. Studies show that vitamin E has preservative function, which prevents the oxidation of chemical compounds in moisturizers.

**Notes**

Based on industry-sponsored studies, the Cosmetic Ingredient Review (CIR) Expert Panel concluded that there are some safety tips in using products containing the AHAs glycolic and lactic acid: the AHA concentration should be 10 percent or less; the final product should have a pH of 3.5 or greater; and the final product should be formulated in such a way that it protects the skin from increased sun sensitivity or its package directions tell consumers to use daily protection from the sun.

## Further Listening

### *Petroleum Jelly*

*Listen to the audio clip and choose the correct answer to each of the following questions.*

**Words and Expressions**

ailment *n.* 疾病，病症      petroleum jelly *n.* 石油冻      odorless *adj.* 无味的
versatile *adj.* 多用途的      insulate *v.* 隔绝，保温      eczema *n.* 湿疹
chapped *adj.* 皲裂的      soothe *v.* 镇静，舒缓      nares *n.* 鼻孔

1. Which of the following is NOT an advantage of petroleum jelly?
   A) It is easy to find.          B) It is inexpensive.          C) It has a pleasant smell.

2. How does petroleum jelly prevent the skin from losing moisture?
   A) It insulates the skin, so it does not lose heat.
   B) It draws water from the environment to add moisture to the skin.
   C) It refreshes the skin by accelerating skin turnover.

3. Why is petroleum jelly regarded as the best way to reduce the risk of eczema in newborns?
   A) Because it protects the newborns from sun damage.
   B) Because it is non-irritant and non-allergenic.
   C) Because it is chemically similar to proteins in human skin.

4. According to Dr. Davis, where should petroleum jelly NEVER be used?
   A) On the fingertips of infants and small children.
   B) In the nose of infants and small children.
   C) On the lips of infants and small children.

## Speaking

### *Section 1   Moisturizer Formulations*

*Read the paragraph below and interpret it into Chinese.*

　　Most moisturizers are combination cosmetic products in which individual ingredients are combined to elicit the desired effect. The aesthetic qualities of the moisturizer are essential for patient and consumer acceptance and compliance as well as for the individual therapeutic design of the moisturizer. Consumers and patients seem to prefer a less greasy product that is applied smoothly. Many moisturizers have an easily spreading, less viscous emollient in combination with a medium spreading, more viscous emollient. A sensation of early, easy smoothness to the skin occurs with thinner, easier-spreading agents, whereas the long-lasting efficacy of the moisturizer is maintained by the thicker components that are more difficult to spread.

### Section 2　Think and Discuss

*Work in a group. Discuss the following questions and share your answers.*

1. What are the common symptoms of both temporary and seasonal dry skin? When is it necessary for one with dry skin concerns to see a doctor?
2. Are there any tips in applying moisturizers for the best effect? How often should one use moisturizers?

### Critical Thinking

#### Project 1　History of Moisturizers

The use of moisturizers by mankind has historic roots. Ancient Egyptians frequently anointed their bodies with oils. The Bible describes applications of oils to the skin, and Ancient Greek and Roman cultures regularly applied oil-containing products. Humans have recognized the value of externally applied lipids for thousands of years, and we continue to value these products. What moisturizers do ancient Chinese use? How do they make moisturizing products? Do the ancient ingredients and formulations still have significance in modern times?

#### Project 2　Safety Concerns of Ingredients in Moisturizers

Moisturizers are usually free from strong irritants; however, repeated exposure of sensitive areas to mildly irritating preparations may cause various skin reactions, including dryness, redness, burning sensation, and itching. Strong irritants are typically easy to identify; however, weak irritants are less obvious to avoid. What ingredients for moisturizers may cause safety concerns? What are their side effects?

### Glossary

1. active ingredients 活性成分
2. amino acid *n.* 氨基酸
3. cell culture 细胞培养物
4. cell proliferation 细胞增殖，细胞增生
5. cocoa butter *n.* 可可脂，可可油
6. collagen synthesis 胶原蛋白合成
7. concentration /ˌkɒnsnˈtreɪʃn/ *n.* 浓度
8. corneocyte /kɔːniːəʊˈsaɪt/ *n.* 角化细胞，角层细胞
9. derivative /dɪˈrɪvətɪv/ *n.* 衍生物
10. desquamation /ˌdeskwəˈmeɪʃən/ *n.* 皮肤脱屑，剥落
11. emollient /iˈmɒliənt/ *n.* 润肤剂
12. epidermal water loss 经表皮失水

13. filaggrin /fɪləgˈrɪn/ n. 丝聚蛋白

14. fine lines 细纹

15. hand sanitizer 消毒洗手液，免水洗手液

16. humectant /hjuːˈmektənt/ n. 保湿剂

17. hydration /haɪˈdreɪʃən/ n. 补水，水合作用

18. hydrocarbon /ˌhaɪdrəˈkɑːbən/ n. 碳氢化合物，烃类，碳氢

19. hydrolytic enzyme 水解酶

20. hydroscopic /ˌhaɪdrəsˈkɒpɪk/ adj. 吸水的

21. inorganic salts 无机盐类

22. intercellular lipids 细胞间脂质

23. ionic interaction 离子作用

24. lactic acid n. 乳酸，2-羟基丙酸

25. lipid-soluble /ˈlɪpɪd ˈsɒljəbl/ adj. 脂溶性的

26. membrane /ˈmembreɪn/ n.（身体内的）膜，（动植物的）细胞膜

27. mineral oil 矿物油，矿油

28. moisturizer /ˈmɔɪstʃəraɪzə(r)/ n. 润肤霜，润肤膏

29. monotherapy /mɒnəʊˈθerəpɪ/ n. 单一治疗，单药治疗

30. NMF (natural moisturizing factor) 天然保湿因子

31. occlusive /əˈkluːsiv/ n. 封闭剂

32. oxidation /ˌɒksɪˈdeɪʃn/ n. 氧化（作用）

33. paraffin /ˈpærəfɪn/ n. 石蜡，煤油

34. PCA (pyrrolidonecarboxylate) 吡咯烷酮羧酸钠

35. petrolatum /ˌpetrəˈleitəm/ n. 矿脂，石油冻，凡士林

36. plasticity /plæˈstɪsəti/ n. 弹性，塑性

37. retinoid /ˈretinɔid/ n. 类视黄醇，维甲酸，维A酸

38. scavenger /ˈskævɪndʒə(r)/ n. 清除剂

39. sensitive reaction 过敏反应

40. skin protectant 皮肤保护剂

41. skin rejuvenator 嫩肤剂

42. sorbitol /ˈsɔːbɪtəl/ n. 山梨醇，山梨糖醇

43. squalene /ˈskwɒliːn/ n. 角鲨烯，三十碳六烯

44. SC (stratum corneum) 角质层

45. urea /juˈriːə/ n. 尿素

46. urocanic acid 尿刊酸，咪唑丙烯酸

47. water-binding capacity 保水力，持水力

8.1　Keys and Scripts

## Lesson 2 Anti-aging

### Checklist for Students

**Key Concepts**

| | | | | |
|---|---|---|---|---|
| herbal extract | antioxidant | photoaging | oxidation | tissue damage |
| free radical | collagen | elastin | hyaluronic | acid |
| resveratrol | lipoic acid | ferulic acid | coenzyme Q10 | signal peptide |
| neurotransmitter-affecting peptide | | carrier peptide | biological peptide | |
| retinoid | receptor | retinol | retinaldehyde | |

**Learning Objectives**

- Learn and understand the major ingredients added in anti-aging products.
- Learn and explain the anti-aging effects of herbal extracts.
- Learn and explain the functions and working mechanism of antioxidants.
- Learn and explain the functions and working mechanism of peptides.
- Learn and explain the functions and working mechanism of retinoids.

### Before Listening

Moisturizing alone can improve the appearance of skin by temporarily plumping the skin, making lines and wrinkles less visible. Moisturizers are lotions, creams, gels and serums made of water, oils and other ingredients, such as proteins, glycerin, lactate and urea. Wrinkle creams often are moisturizers with active ingredients that offer additional benefits. They include herbal extracts, antioxidants, proteins and peptides, retinoids, hydroxy acids, and sunscreens, which are claimed to prevent oxidative reactions and the formation of free radicals, resurface the epidermis, and promote the natural synthesis of collagen and elastin. These added ingredients are intended to improve skin tone, texture, fine lines and wrinkles.

### Listening

*Section 1    Herbal Extracts*
*Listen to the audio clip and tell if the following statements are true (T) or false (F).*

1. Herbal extracts are often used as bioactive agents in anti-aging products.
2. Coffee and pomegranate have soothing effect for skin.

8.2  Section 1

116

3. Black tea and olive have anti-inflammatory effects.

4. Chamomile and mushrooms are natural antioxidants.

5. Blueberry and ginseng can restore damaged cells.

6. Peppermint and jojoba extracts have emolliating effect.

7. Herbal extracts are often used in combination for enhanced benefits.

8. Herbal extracts are used in large amounts in cosmetic products for the best therapeutic effects.

9. Certain ingredients, like green tea, are claimed to influence skin mechanism in a way that is beneficial for anti-aging.

10. The effectiveness of herbal extracts has been fully validated by clinical studies.

### Section 2   *Antioxidants*

*Listen to the audio clip and fill in the blanks with the missing information.*

8.2   Section 2

1) _____ is believed to be the main reason for both chronological aging and photoaging. Even though the human skin is able to mitigate the negative effects of oxidation with its efficient antioxidant system, this intrinsic network fails to 2) _____, which leads to 3) _____ _____. The free radical theory of aging believes that oxidative stress increases with age, 4) _____. Free radicals that are developed from oxidation will 5) _____, causing signs of aging. Therefore, to slow down the aging process, we need to supply the skin with antioxidants, which can protect the cells from aging by 6) _____. Antioxidants are widely used, not only in an increasing number of cosmetic and OTC cosmeceuticals, but also in drinks and foods. The most common ingredients are 7) _____. Other ingredients are also used, such as green tea and grape-seed extract; resveratrol, lipoic acid, ferulic acid, and coenzyme Q10. It is not yet known which antioxidants have the best effects in preventing photoaging, but it is always recommended that 8) _____ should be taken orally together with topical products. Antioxidants can not only serve as protective molecules against oxidation. Some of them can also 9) _____, preserve hyaluronic acid levels in the skin, and also 10) _____.

---

**Notes**

---

Coenzyme Q10 is an obligatory member of the respiratory chain in the mitochondria of all cells. Therefore, it is an essential ingredient in the formation of adenosine triphosphate (ATP), the source of energy in most cellular processes. Coenzyme Q10 provides an antioxidant action either by direct reaction with free radicals or by regeneration of tocopherol and ascorbate from their oxidized state.

8.2    Section 3

### *Section 3    Proteins and Peptides*

*Listen to the audio clip. Match the descriptions with the categories of peptides.*

A) They act as messengers that activate fibroblast cells to produce more collagen or reduce the decomposition of existing collagen.

B) They act as antioxidants for aging by transferring hydrogen atoms and single electrons and chelating the pro-oxidative metals transition.

C) They can stabilize and provide essential trace elements for enzymatic processes.

D) They are designed to simulate botulinum toxin with the principle of blocking acetylcholine release at the neuromuscular junction, reduce facial muscle contraction, and decrease sagging and skin wrinkling following topical application.

E) In using this type of peptides, the specific chains of amino acids could stimulate the growth of human skin dermal fibroblasts in vitro, and reduce rough and coarse wrinkles in human subjects.

F) They have properties of stabilizing and transporting copper into cells. Copper is considered a fundamental metal responsible for improving skin oxidation, enzymatic reactions, and angiogenesis.

G) These peptides aim to raise the threshold for small muscle activities with required and extra signals. These signals are able to decrease subconscious muscle movement in time for muscle activity.

1. _____ Signal peptides
2. _____ Neurotransmitter-affecting peptides
3. _____ Carrier peptides
4. _____ Biological peptides

---

**Notes**

Known as "pentapeptides", anti-aging peptides are often included in skincare products to help promote anti-aging effects in individuals who wish to combat the aging process. These peptides actually help to form collagen. As such, they may display powerful regenerative properties. Collagen is a form of protein and is the building block necessary to develop almost every "tissue" based component of our body from our skin to our vital organs, and many other elements in between.

---

8.2    Section 4

### *Section 4    Retinoids*

*Listen to the audio clip and choose the best answer for each of the following questions.*

1. Which of the following is NOT true about Tretinoin?

   A) It is the first retinoid under the brand name Retin-A.

   B) It was originally used as an acne treatment in the 1970s.

   C) It can lead to improvement and changes in the epidermis.

2. How do retinoids affect skin cells?

A) They bind to the receptors within the cells' nucleus.

B) They transmit nutrients into the cells.

C) They help the skin to hold moisture in the cells.

3. How do retinoids reduce fine lines and wrinkles?

A) They reduce fine lines and winkles by stimulating contraction of blood vessels.

B) They reduce fines lines and wrinkles by increasing the production of collagen.

C) They reduce fine lines and wrinkles by lightening up the skin tone.

4. What is the advantage of retinol compared with retinoic acid in low concentrations?

A) It is safer and more affordable.

B) It is more effective on intrinsic skin aging.

C) It is better tolerated for the users.

5. Why is it necessary to wear sunscreen during the use of retinoids?

A) Because retinoids can cause skin dryness and irritation.

B) Because it takes a regular use for improvements to be apparent.

C) Because retinoids increase skin sensitivity to sunlight.

## Further Listening

8.2    Further Listening

***Anti-aging Foods that Will Make Your Skin Glow***

*Listen to the audio clip and complete the chart.*

### Words and Expressions

| | | |
|---|---|---|
| ellagic acid *n.* 鞣花酸 | punicalagin *n.* 安石榴苷 | carotenoid *n.* 类胡萝卜素 |
| lycopene *n.* 番茄红素 | lutein *n.* 叶黄素 | probiotics *n.* 益生菌 |
| lactic acid *n.* 乳酸 | turmeric *n.* 姜黄粉 | resveratrol *n.* 白黎芦醇 |
| magnesium *n.* 镁 | manganese *n.* 锰 | polyphenol *n.* 多酚 |

| Foods | Functions |
|---|---|
| Oranges | |
| Pomegranates | |

(continued)

| Foods | Functions |
|---|---|
| Tomatoes | |
| Spinach, kale | |
| Belle pepper | |
| Fermented foods | |
| Turmeric | |
| Raw cacao | |
| Oats | |
| Olive oil | |

## Speaking

### Section 1    Anti-wrinkle Regimen
*Read the paragraph below and interpret it into Chinese.*

An anti-wrinkle cream may lessen the appearance of your wrinkles, depending on how often you use it, the type and amount of active ingredients in the wrinkle cream, and the type of wrinkles you want to treat. In addition to using anti-aging products, there are also more reliable ways to improve and maintain your skin's appearance: first, protect your skin from the sun. Exposure to UV light speeds up the natural aging process of your skin, causing wrinkles and rough, blotchy skin. In fact, sun exposure is the main reason for signs of aging in the skin, including uneven pigmentation. It is always necessary to limit the time of sun exposure and wear sunscreen and protective clothing to prevent future wrinkles. Moisturizers can't prevent wrinkles, but they trap moisture in the skin, temporarily masking fine lines and creases. Smoking causes narrowing of the blood vessels in the outermost layers of your skin. It also damages collagen and elastin — fibers that give your skin its strength and elasticity. As a result, skin begins to sag and wrinkle prematurely. Even if you've smoked for years or smoked heavily, you can still improve your skin tone and texture and prevent future wrinkles by quitting smoking.

*Section 2　Think and Discuss*
*Work in a group. Discuss the following questions and share your answers.*

1. When do you think is the age threshold for using anti-aging products? Why?
2. Can the signs of aging skin be reversed? Why or why not?

## Critical Thinking

*Project 1　Medical Treatments for Wrinkles*

A dermatologist can help you create a personalized skin care plan by assessing your skin type, evaluating your skin's condition and recommending products likely to be effective. If you're looking for more dramatic results, you can look for medical treatments for wrinkles, including prescription creams, botulinum toxin (Botox) injections or skin-resurfacing techniques. What are the advantages and disadvantages of such treatments? How effective are they? Carry out a research on this topic.

*Project 2　Chemical Peeling*

Chemical peeling is a minimally invasive procedure used for the cosmetic improvement of the skin. During this procedure, a chemical agent of a defined strength is applied to the skin, which causes a controlled destruction of the skin layers with subsequent regeneration and rejuvenation of the tissues, resulting in improvement of texture and surface abnormality. What are the main categories of chemical peeling? How is it operated and what tips do people need to pay attention to in chemical peeling?

## Glossary

1. acetylcholine /ˌæsɪtɪlˈkɒliːn/ *n.* 乙酰胆碱
2. angiogenesis /ˌændʒiəuˈdʒenisis/ *n.* 血管生成
3. bioactive agents 生物活性成分
4. biological peptides 生物活性肽
5. botulinum toxin 肉毒杆菌毒素
6. carrier peptides 载体胜肽
7. chamomile /ˈkæməmaɪl/ *n.* 甘菊，黄春菊
8. chronological aging 时间性老化，自然老化

9. coenzyme Q10 辅酶 Q10
10. connective tissue 结缔组织
11. copper /ˈkɒpə(r)/ *n.* 铜
12. cosmeceutical /kɒzmiːsˈjuːtɪkl/ *n.* 功能性化妆品
13. electron /ɪˈlektrɒn/ *n.* 电子
14. enzymatic process 酶促过程
15. extrinsic aging 外源性老化

16. facial muscle contraction 面部肌肉收缩

17. ferulic acid 阿魏酸，4-羟-3-甲氧肉桂酸

18. fibroblast cells 纤维母细胞

19. ginseng /ˈdʒɪnseŋ/ n. 人参，西洋参

20. herbal extracts 植物提取物

21. hydrogen atom 氢原子

22. in vitro adj. 体外的，在试管内的

23. intrinsic skin aging 内源性皮肤老化

24. lipoic acid 硫辛酸

25. neuromuscular junction 神经肌肉接头

26. neurotransmitter-affecting peptides 神经传导二胜肽

27. pharmaceutical /ˌfɑːməˈsuːtɪkl/ adj. 制药的

28. photoaging /fəʊtəʊˈeɪdʒɪŋ/ n. 光老化

29. physicochemical /fɪzɪkəʊˈkemɪkəl/ adj. 物理化学的

30. pro-oxidantive /ˈprəʊɒksɪˈdæntɪv/ adj. 助氧化的

31. receptor /rɪˈseptə(r)/ n. 感受器，接收器

32. regimen /ˈredʒɪmən/ n. (护肤)方案，养生之道

33. resveratrol /rezˈvɪərɪˌtrɔːl/ n. 白藜芦醇

34. retinaldehyde /ˈretɪnldɪhaɪd/ n. 视黄醛，A醛

35. retinol /ˈretɪnɒl/ n. 视黄醇

36. sagging /ˈsægɪŋ/ n. 下垂，松垂

37. signal peptides 信号肽

38. synthetic /sɪnˈθetɪk/ adj. 合成的，人造的

39. topical application n. 局部施用

40. trace elements 痕量元素，微量元素

8.2 Keys and Scripts

---

## Lesson 3　Anti-acne

### Checklist for Students

#### Key Concepts

| | | | | |
|---|---|---|---|---|
| acne | comedone | P. acne | pore | bacteria |
| hormonal imbalance | benzoyl peroxide | salicylic acid | antimicrobial | anti-inflammatory |
| tretinoin | Retin-A | Renova | Retin-A Micro | Tazorac |
| Avita | azelaic acid | | | |

#### Learning Objectives
- Have a basic understanding of acne.
- Learn and understand benzoyl peroxide treatment for acne.

- Learn and understand exfoliating preparation and its effect on acne.
- Learn and understand anti-acne topical treatments of retinoids and azelaic acid.
- Understand ingredients to avoid in treating acne.

## Before Listening

Acne is a skin disease often characterized by inflammation in the skin, erupting in lesions. This condition is widespread, particularly during adolescence, and the majority of people experience it at some point during their life. Acne is officially called acne vulgaris, which is caused by multiple factors. Hormonal imbalances in teenagers are a major reason for acne. Most of the problems come from bacteria on the skin that start to grow in abundance. Most of these bacteria grow in clogged pores that are full of dead skin cells. Acne can make people feel unhappy about their appearance, develop poor self-esteem and not want to be around others.

## Listening

### Section 1   Benzoyl Peroxide Treatment for Acne
*Listen to the audio clip and fill in the blanks with the missing information.*

8.3   Section 1

Benzoyl peroxide was first used to treat acne in 1965 and since then has been increasingly relied on to treat acne, either alone or in combination with other anti-acne medications, such as retinoids. When used to treat acne today, benzoyl peroxide is found in 1) _____ at concentrations ranging from 2) _____ . It works in four ways to treat acne.

First, benzoyl peroxide quickly absorbs into the skin where it 3) _____ almost immediately. In this way, benzoyl peroxide may 4) _____ . Benzoyl peroxide also functions as a 5) _____ by drying the skin and helping skin cells flake off. This helps to 6) _____ and leading to acne lesions. Finally, benzoyl peroxide 7) _____ . This is important because skin pores are filled with skin oil, and benzoyl peroxide can therefore penetrate into pores where it can work to kill bacteria and unclog the pore.

Effective as it is in treating acne, there are side effects associated with benzoyl peroxide, which include 8) _____ . These side effects occur within the first few weeks of treatment, and dramatically subside after a few weeks of use. When using benzoyl peroxide, it's important to start slowly, 9) _____ of benzoyl peroxide and only using it once a day for the first week to limit these side effects. Then, you can move to 10) _____ _____ and slowly increase your dosage of benzoyl peroxide so that after one month a generous amount of benzoyl peroxide can be used to adequately clear the skin.

**Notes**

When it comes to using benzoyl peroxide during pregnancy, the US Food and Drug Administration (FDA) has classified it as a Category C drug, which means that its effects on the health of the fetus are unknown. However, doctors generally consider it safe to use during pregnancy because very little of it is absorbed into the blood.

### Section 2   Salicylic Acid Treatment for Acne

*Listen to the audio clip and complete the following chart.*

8.3   Section 2

📖 **Overview**

   *Salicylic acid, or beta hydroxy acid (BHA), can be used to treat many skin problems because it has multiple functions. When combined with benzoyl peroxide, it is especially effective for many systemic causes of blemishes.*

| Properties | Functions | Advantages |
|---|---|---|
| anti-inflammatory | derivative of 1) _____ reduce 2) _____ | diminish and eliminate redness and blemishes, 3) help _____, prevent _____, and decrease _____ |
| exfoliating | improve the shape of 4) _____ exfoliate inside 5) _____ 6) dislodge _____ | lipid soluble: 7) _____ anti-inflammatory: 8) _____ |
| antimicrobial | kill bacteria | prevent and cure acne |

### Section 3   Tretinoin Treatment for Acne

*Listen to the audio clip and give short answers to the questions below.*

8.3   Section 3

📖 **Overview**

   *Tretinoin is a form of vitamin A and, therefore, comes under the general heading of retinoids. The best-known products that contain tretinoins are Retin-A, Renova, Retin-A Micro, Tazorac, Avita, and generic tretinoin. These are all basic treatments for blemishes because they change the way skin cells are formed in the layers of skin as well as in the pore.*

1. What will happen if skin cells have an abnormal shape?

_____

2. What is the main effect of tretinoin in treating acne?

_____

_____

_____

3. Why should the application of tretinoin and benzoyl peroxide be separated? What is the recommended application?

_____

_____

4. What are the usual concentrations of tretinoin in treating acne?

_____

5. What are the side effects of using tretinoin preparations?

_____

_____

### *Section 4    Azelaic Acid Treatment for Acne*

*Listen to the audio clip and choose the correct answer to each of the following questions.*

8.3    Section 4

1. What kind of substance is azelaic acid?
   A) It's a substance chemically synthesized in the lab.
   B) It's a substance made naturally in the body.
   C) It's a substance extracted from azalea.
2. Which of the following is NOT a therapeutic effect of azelaic acid?
   A) It is anti-bacterial and anti-inflammatory.
   B) It reduces sebum secretion.
   C) It regulates the cell turnover within the follicle.
3. How effective is azelaic acid in treating acne according to the two randomized, controlled clinical trials?
   A) There was a 70% to 71% reduction of facial papules and pustules.
   B) There was a 77% reduction of facial papules and pustules.
   C) There was a 63% reduction of facial papules and pustules.
4. When can the efficacy of azelaic acid be enhanced?
   A) When it is used in monotherapy.
   B) When it is used with a concentration of over 25%.
   C) When it is combined with oral antibiotics.
5. Which of the following is NOT a possible side effect of azelaic acid?
   A) Sun sensitivity.
   B) Skin dryness.
   C) Lightening of the skin where applied.

**Notes**

Azelaic acid, also called nonanedioic acid, is an organic compound with the formula $(CH_2)_7(CO_2H)_2$, and the best-known dicarboxylic acid. Its name stems from the action of nitric acid (azote, nitrogen, or azotic, nitric) oxidation of oleic or elaidic acid: az means "azote" or "nitrogen", and elaion means "oil" in Latin, hence "azelaic acid". It is translated into "壬二酸" in Chinese, as it has 9 carbon atoms. According to the naming principle of chemical compounds in Chinese, the numbers are represented in "天干地支"(the Ten Heavenly Stems and Twelve Terrestrial Branches), and the number 9, or its English word roots "nonyl-", "nona-" match with the Chinese "壬", or "九".

## Further Listening

8.3 Further Listening

***Why You Should Never Pop the Pimples***
*Listen to the audio clip and tell if the following statements are true (T) or false (F).*

## Words and Expressions

| | |
|---|---|
| non-medicated *adj.* 非药用的 | non-comedogenic *adj.* 不会引发黑头的 |
| sink *n.* 洗脸池 | dispense *v.* 挤出 |
| lather *n.* 泡沫 | dab *v.* 轻拍 |
| subside *v.* 消退 | smear *v.* 涂抹 |

1. Popping the white heads can lead to more sebum production.
2. Pimples are bulging bags of oil and bacteria that cannot be released.
3. One has the most sebaceous glands on the face and hand.
4. Squeezing pimples can result in excess oil being pushed deeper into the dermis.
5. Squeezing will make the inflammation worse and lead to darker skin and scars.
6. Pimples usually resolve over two weeks without scaring.

## Speaking

*Section 1   Irritating Ingredients that Aggravate Acne*
*Read the paragraph below and interpret it into Chinese.*

Many blemish products available on the market contain ingredients that make the breakouts even worse, or cause more skin problems. Such ingredients include harsh surfactants, abrasive scrub particles, alcohol, menthol, peppermint, camphor and grapefruit oil. The main problem with these ingredients is that they are highly irritating, which can impair the skin barrier against bacteria and stimulate more sebum production. Another concern is that such ingredients, though claimed to "fight blemish", actually have no effect in reducing oil surplus, help exfoliation or keep hormonal balance. By killing more skin cells than necessary, they cause further irritation, skin redness and dryness, and clogged-up pores. Therefore, skin care products such as facial masks, astringents, toners and facial scrubs with irritating ingredients should be avoided, as they aggravate acne. What is worse, they can even lead to small pimples. Another type of product to avoid is bar soaps, because they contain ingredients that clog the pore, such as tallow and wax-based thickening agents.

*Section 2   Think and Discuss*
*Work in a group. Discuss the following questions and share your answers.*

1. How to treat scars and blemishes left by acnes on the skin?
2. What treatments and cosmetic products would you recommend to people with mild, moderate and severe acne respectively?

## Critical Thinking

*Project 1   Psychological Impact of Acne*

Acne infection may leave on the skin a permanent physical scarring. Acne has a negative psychological effect on the victims: it can lower self-esteem, allow negative mood to set in, and even cause a higher risk of anxiety, depression and suicidal tendencies in the victims. Analyze the emotional, mental, social and psychological effects of acne on individuals and present your findings.

*Project 2   Acne Myths*

Sometimes when you look for acne treatments, you often come across many myths associated with acne. The most prevalent one is eating chocolate. Another myth related to it is eating greasy food. The truth is there is no evidence that the food you eat can cause acne. Myth also has that dirty skin causes acne. Unfortunately, blackheads and other kinds of lesions from acne aren't caused by dirty skin. Do you know any other acne myths? How do the myths form and spread? Conduct a research on this issue.

## Glossary

1. abrasive scrub particles 研磨颗粒
2. antibiotics /ˌæntɪbaɪˈɒtɪks/ n. 抗生素
3. antimicrobial /ˌæntɪmaɪˈkrəʊbɪəl/ adj. 抗菌的 n. 抗菌剂
4. aspirin /ˈæsprɪn/ n. 阿司匹林
5. Avita 维生素 A
6. azelaic acid 壬二酸
7. barley /ˈbɑːli/ n. 大麦
8. benzoyl peroxide 过氧化苯甲酰
9. blemish /ˈblemɪʃ/ n. 斑点，疤痕，瑕疵
10. camphor /ˈkæmfə(r)/ n. 樟脑
11. clindamycin /klɪndəˈmaɪsɪn/ n. 克林霉素
12. comedones /ˌkɒmɪˈdəʊniːz/ n. 黑头粉刺
13. dicarboxylic acid n. 二羟酸
14. dislodge /dɪsˈlɒdʒ/ v. 驱逐，逐出
15. dosage /ˈdəʊsɪdʒ/ n.（通常指药的）剂量
16. exfoliant /eksˈfəʊliənt/ n. 去角质成分，去角质剂
17. inflammation /ˌɪnfləˈmeɪʃn/ n. 发炎，炎症
18. lesion /ˈliːʒn/ n.（因伤病导致皮肤或器官的）损伤，损害
19. P. acnes (propionibacterium acnes) 痤疮丙酸杆菌
20. papule /ˈpæpjuːl/ n. 丘疹，小突起
21. peeling agent 剥离剂
22. pore /pɔː(r)/ n. 毛孔
23. pustule /ˈpʌstjuːl/ n. 脓疱
24. Renova 维 A 酸润肤霜
25. Retin-A micro 维甲酸微球
26. Retin-A 全反维 A 酸
27. retinoids /rɪtɪˈnɔɪdz/ n. 类视黄醇，类视黄醇类物质
28. rye /raɪ/ n. 黑麦
29. salicylic acid n. 水杨酸
30. scarring /ˈskɑːrɪŋ/ v. 给...留下疤痕 n. 瘢痕形成
31. sebum /ˈsiːbəm/ n. 皮脂
32. subside /səbˈsaɪd/ v. 消退，减弱
33. swelling /ˈswelɪŋ/ n. 肿胀，膨胀
34. Tazorac 维 A 酸产品罗肤格
35. tretinoin /trətɪnˈɔɪn/ n. 维甲酸，维 A 酸
36. unclog /ˌʌnˈklɒg/ v. 清除油污（堵塞物）
37. yeast /jiːst/ n. 酵母，酵母菌

8.3 Keys and Scripts

# Unit 9  Products for Special Skin Concerns

## Checklist for Students

### Key Concepts

| | | | |
|---|---|---|---|
| UVA | UVB | SPF | sunscreen |
| after-sun product | hydration therapy | physical sunscreen | chemical sunscreen |
| broad-spectrum | protection | waterproofing agent | photostabilizer |
| film-forming | ingredient | gels | tan |
| stick | emulsion | aerosol spray | ointments and oil |

### Learning Objectives

● Learn the types and definitions of sun care products.
● List the main ingredients of sunscreens and explain their respective functions.
● List and describe different product forms of sunscreens.
● Learn and break the myths related to sun protection.
● Understand the importance of sunscreen and how to choose specific products.

## Before Listening

Exposure to sunlight enables the body to produce vitamin D, which plays a crucial role in skeletal development, immune function and blood cell formation. While some sunlight is beneficial and its light and warmth enhance people's general feeling of well-being, prolonged exposure to UV radiation can cause serious health damages, which include sunburn, skin cancer, skin aging, etc. Sunscreen protects your skin from the sun's ultraviolet, or UV light. There are two types of UV light that can harm your skin: UVA and UVB. UVA is the long wavelength of light that penetrates to the deep layers of skin. UVB is the shorter wavelength of light that penetrates the surface of the skin and causes sunburn. The best sunscreens offer protection from all UV light. These are labeled as "broad-spectrum" or "full-spectrum" sunscreens.

## Listening

### *Section 1　Types and Definitions of Sun Care Products*

*Listen to the audio clip and fill in the blanks with missing information.*

9.1　Section 1

Sun exposure is unavoidable in our daily lives; however, without appropriate protection, the sun can severely damage our skin in the short as well as long term. Sunscreens are considered 1) ＿＿＿＿＿＿＿＿＿＿＿＿＿ in the US since they help prevent various skin conditions, including sunburn, aging, and skin cancer. After-sun products, on the other hand, are considered 2) ＿＿＿＿＿＿＿＿ in the US. There are two basic types of sunscreens on the market: physical sunscreen and chemical sunscreen. Physical sunscreen literally 3) ＿＿＿＿＿＿＿＿＿ from reaching your skin by either 4) ＿＿＿＿＿＿＿＿＿＿＿. These products contain zinc oxide or titanium dioxide. Some people don't like to use physical sunscreen, because it remains visible on the skin after you apply it. It also can be hard to wash off. However, physical sunscreen provides 5) ＿＿＿＿＿＿＿＿＿＿＿＿＿＿＿. It also tends to be less irritating to the skin than chemical sunscreens. That can be particularly beneficial for 6) ＿＿＿＿＿＿＿＿＿＿＿＿＿＿. Chemical sunscreens work by absorbing the UV light and causing it to 7) ＿＿＿＿＿＿＿＿＿＿＿＿＿＿ that prevents it from damaging your skin. Both of these types are rated with a sun protection factor (SPF), which lets the consumer know 8) ＿＿＿＿＿＿＿＿＿＿＿＿＿＿ the product provides. The SPF of a product is the 9) ＿＿＿＿＿＿＿＿＿＿＿ required for a person's protected skin to redden after being exposed to sunlight compared to the time required for the same person's 10) ＿＿＿＿＿＿＿＿＿＿ to redden. For example, a product with SPF15 means that a person whose unprotected skin would redden in ten minutes can apply the product and stay in the sun 11) ＿＿＿＿＿＿＿＿＿＿, or 150 minutes, before they get a sunburn. After-sun products are designed to be used after 12) ＿＿＿＿＿＿＿＿ or other UV radiation. After sunbathing, even without any signs of 13) ＿＿＿＿＿＿, appropriate skin care is recommended. After-sun preparations help 14) ＿＿＿＿＿＿＿＿＿＿ the skin as well as provide 15) ＿＿＿＿＿＿ and relieve pain resulting from sunburn. Product forms include lotions, creams, and gels.

**Notes**

Carbon-based chemical sunscreens can harm marine life. Take coral reefs, for example. Although they cover less than 1% of the Earth's underwater surface, they're home to nearly 25% of all fish species, making them the most diverse and productive marine ecosystems. Research shows that carbon-based chemical sunscreen ingredients, like oxybenzone, butylparaben, octinoxate, and 4MBC contribute to a stress condition called coral bleaching in corals, which are living creatures. Exposure to these organic compounds results in the death of the coral's symbiotic algae.

## *Section 2    Main Ingredients of Sunscreen Products*

*Listen to the audio clip and match the functions with the ingredients.*

9.1    Section 2

| | |
|---|---|
| A. inorganic UV filters | B. organic UV filters |
| C. waterproofing agents | D. photostabilizers |
| E. emollients | F. film-forming ingredients |
| G. antioxidants | |

1. _____ They mainly include two types: titanium dioxide and zinc oxide.
2. _____ They prevent the degradation of organic UV filters.
3. _____ They are added to sunscreens to help prevent oxidative reactions.
4. _____ They are generally aromatic compounds with a molecular structure responsible for absorbing UV energy.
5. _____ They are added to increase sunscreen's water-resistance properties.
6. _____ They are ingredients that help form an even and uniform film on the skin after application and drying.
7. _____ They are lipophilic ingredients that can increase water resistance and serve as solvents for lipophilic organic sunscreens.
8. _____ They convert the absorbed energy into longer, lower energy wavelengths.

**Notes**

Sun blocking agents such as titanium dioxide and zinc oxide have been found to be highly protective against ultraviolet, visible, and even infrared radiation. They are also photostable. However, these mineral sunscreens tend to be white or opaque on the skin and therefore cosmetically unacceptable. This can be overcome by micronization to decrease particle size and produce particles that are transparent to visible light and appear natural and invisible on the skin.

## *Section 3    Product Forms of Sunscreen*

*Listen to the audio clip and give short answers to the questions below.*

9.1    Section 3

🖳 **Overview**

*Sunscreens are available in a variety of dosage forms, including O/W or W/O emulsions and anhydrous systems, such as ointments, sticks, oils, and silicone-based aerosols, wipes, and gels.*

1. What are the advantages of emulsion sunscreens?

_____

_____

2. What are stick sunscreens suitable for?

_____

_____

3. What is the benefit of multi-position aerosol sunscreen sprays?

_____

_____

4. What are the demerits of ointments and oils?

_____

_____

5. What are the pros and cons of gel sunscreens?

_____

_____

6. What do formulators place on sunscreen wipes to offer additional benefits?

_____

_____

### Section 4   Sun Protection Myths

*Listen to the audio clip and tell if the following statements are true (T) or false (F).*

9.1   Section 4

1. Higher levels of melanin in the skin will absorb more UVB rays.
2. Water- or sweat-resistant sunscreens will not wear off when exposed to water, and therefore do not need to be reapplied.
3. Most sunscreens have varying SPFs and different active ingredients, varying from physical ones to chemical ones like avobenzone.
4. Experts advise that certain amount of vitamin D can still be absorbed after 5–30 minutes of sun exposure per day.
5. Sunscreen is only necessary on sunny days, or when the skin is exposed.

> **Notes**
>
> Avobenzone is a sunscreen agent that protects against the full spectrum of UV light. Of all sunscreen agents, avobenzone has one of the largest absorbance spectrums, absorbing light between 320–400 nm (peak absorption ∼ 360 nm). Avobenzone is susceptible to photodegredation, and therefore it is important that avobenzone be combined with photostabilizers in the final sunscreen product.

9.1　Further Listening

## Further Listening

***Why Do We Wear Sunscreens?***
*Listen to the audio clip and choose the correct answer to each of the following questions.*

### Words and Expressions

| | | |
|---|---|---|
| ultraviolet rays 紫外线 | wavelength *n.* 波长 | absorption *n.* 吸收 |
| chromophore *n.* 发色团 | molecule *n.* 分子 | hemoglobin *n.* 血红蛋白 |
| melanin *n.* 黑色素 | carcinogenesis *n.* 致癌 | zinc oxide 氧化锌 |
| titanium dioxide 二氧化钛 | diaper *n.* 尿布 | deteriorate *v.* 消退 |
| transparent *adj.* 清透的 | allergic *adj.* 过敏的 | immune *adj.* 免受的 |
| melanoma *n.* 黑色素瘤 | mutation *n.* 变异 | aesthetic *adj.* 美观性的 |
| elasticity *n.* 弹性 | saggy *adj.* 下垂的 | chronically *adv.* 慢性地 |
| basal cell carcinoma 基底细胞癌 | squamous cell carcinoma 鳞状细胞癌 | |

1. Which of the following is NOT a harm of UVA?
   A) Sunburn.
   B) Tanning.
   C) Carcinogenesis.
2. What is the purpose of adding zinc oxide in sunscreens?
   A) To better absorb the rays.
   B) To create a physical barrier.
   C) For easier application in using.
3. Which of the following is an advantage of chemical sunscreens?
   A) They reduce users' allergic reactions.
   B) They have better protection against photoaging.
   C) They are more transparent when rubbed on the skin.
4. How does sunscreen protect you from skin cancer?
   A) It blocks ultraviolet rays which lead to mutations in skin cells.
   B) It strengthens our built-in sun protection functions.
   C) It helps the skin to recover sooner from sunburn.
5. What skin problem does photoaging from chronic sun exposure cause?
   A) Irritation.
   B) Loss of elasticity.
   C) Acnes.

## Speaking

### Section 1    Which Sunscreens Are Best for Children?
*Read the paragraph below and interpret it into Chinese.*

Since kids are more vulnerable to sun damage and the harmful effects of chemical exposure, they should use a sunscreen rated highly for safety and one that offers effective protection from UVA and UVB radiation. If your child plans to swim and play in the water, look for a product labeled as providing 80 minutes of water resistance, the highest possible rating. Don't buy sprays or products with bug repellent. Apply sunscreen generously before children go outside and reapply it often. Sunscreen is just one part of a sun-healthy lifestyle. Limiting sun exposure and wearing protective clothing are more important. The American Academy of Pediatrics recommends avoiding sunscreen for children younger than six months, but if you can't find protective shade and clothing, apply a minimal amount of sunscreen to exposed skin, since young children don't yet have protective melanin proteins.

### Section 2    Think and Discuss
*Work in a group. Discuss the following questions and share your answers.*

1. How high an SPF should you use?
2. Will sunscreen protect you from cancer and wrinkles?
3. Is good sunscreen all you need to stay safe in the sun?

## Critical Thinking

### Project 1    How Much Solar Radiation Are We Exposed to?

Many factors can influence how much UV radiation we are exposed to, including both external ones such as geography, and internal ones, such as skin type. Carry out a research on the major factors that influence the amount of solar exposure we have and make a presentation.

### Project 2    Photoaging and Sun Exposure

Aging in humans is a complex, multifactorial process and it is generally admitted that two components play a decisive role: one linked to the genetic background of the individuals and one related to their lifestyles and exposure to environmental mutagens, such as those produced from sun exposure. Explore the relationship between sun exposure and photoaging and present the causal analysis.

## Glossary

1. aerosol spray 喷雾
2. after-sun product 晒后护肤品
3. aromatic /ˌærəˈmætɪk/ *adj.* 芳香的
4. avobenzone /əvəʊˈbenzn/ *n.* 阿伏苯宗
5. broad-spectrum protection 广谱防晒
6. bug repellent 驱虫剂
7. chemical sunscreen 化学防晒剂
8. efficacy /ˈefɪkəsi/ *n.* 功效
9. expiration date 过期时间
10. free radical *n.* 自由基
11. inorganic /ˌɪnɔːrˈgænɪk/ *adj.* 无机的
12. insoluble /ɪnˈsɒljəbl/ *adj.* 不溶的
13. lip balm 润唇膏
14. lipophilic /lɪpəʊˈfɪlɪk/ *adj.* 亲脂的
15. melanin /ˈmelənɪn/ *n.* 黑色素
16. multi-position /ˈmʌlti pəˈzɪʃn/ *adj.* 多位点的
17. ointment /ˈɔɪntmənt/ *n.* 油膏，软膏
18. organic /ɔːrˈgænɪk/ *adj.* 有机的
19. photostabilizer /fəʊtəʊ ˈsteɪbəlaɪzə(r)/ *n.* 光稳定剂
20. physical sunscreen 物理防晒剂
21. reflect /rɪˈflekt/ *v.* 反射（声、光、热）

22. skin cancer 皮肤癌
23. SPF (sun protection factor) 防晒指数
24. SPF booster SPF 促进剂
25. sprayable lotion 喷雾乳液
26. stick sunscreen 防晒棒
27. sunburn /ˈsʌnbɜːn/ *n.* 晒伤
28. susceptible /səˈseptəbl/ *adj.* 容易受影响的
29. tan /tæn/ *v.* 晒黑 *n.* 日晒形成的棕褐肤色
30. titanium dioxide 二氧化钛
31. UV filter 紫外线滤光剂
32. UV light (ultraviolet light) 紫外线
33. UVA (ultraviolet A) 紫外线 A
34. UVB (ultraviolet B) 紫外线 B
35. zinc dioxide /zɪŋk daɪˈɒksaɪd/ 二氧化锌

9.1　Keys and Scripts

## Lesson 2　Hair Styling Products

## Checklist for Students

### Key Concepts

| | | | |
|---|---|---|---|
| permanent waving | perming | reduction | neutralization |
| hair colorant | reducing agent | alkaline pH | temporary dye |

semi-permanent hair dye    progressive hair dye    demi-permanent product
permanent hair coloring    hair bleaching          redyeing        aerosol can
plastic bottle             hair coloring kit

## Learning Objectives

- Understand and describe the process of hair perming.
- List major perming products and explain their differences.
- List different types of hair coloring products and describe their features.
- Understand and describe hair bleach and touch-up.
- Learn the environmental and health concerns of hair styling products.

## Before Listening

As an important part of our life, hair styling can either make temporary changes in the appearance of the hair with various devices, or use permanent styling methods to physically and chemically change the hair into desired shapes. Hair coloring is another popular personal hair care to express individuality and fashion. Across the world, both men and women change their original hair color by either removing the existing hair color or adding a new one.

## Listening

### Section 1    Definition and Process of Hair Perming
*Listen to the audio clip and put the procedures of hair perming into the correct order.*

9.2    Section 1

📖 **Overview**

*A hair perm is a chemical process that changes the hair shape to create a new look. Those with naturally straight hair can transform their hair into waves or curls and those with curly hair can straighten their hair. Some of the different types of products include spiral, stack, spot and volumizing perms. Permanent waving, or hair perm, is usually a two-step process including reduction and neutralization.*

1. Apply a fixing lotion, otherwise known as neutralizing lotion, which contains an oxidizing agent to the hair fibers to break disulfide bonds and oxidize cysteine.
2. Wind the hair tresses on rods that have holes inside to allow the perming lotion to contact all the surfaces of the hair shaft.
3. Wash the hair and section it into smaller areas. Wrap the end of the hair in each area into a thin sheet of tissue paper.
4. Thoroughly rinse the hair with water after unwinding to remove the neutralizer.

**Correct order:** (      ) —— (      ) —— (      ) —— (      )

---

**Notes**

Hair is a biological polymer, with over 90% of its dry weight made of proteins called keratins. Under normal conditions, human hair contains around 10% water, which modifies its mechanical properties considerably. Hair proteins are held together by disulfide bonds from the amino acid cysteine. These links are very robust, and the disulfide links also cause hair to be extremely resistant to protein digestive enzymes. Breaking and making disulfide bonds governs the phenomenon of wavy or frizzy hair. It is breaking and remaking of the disulfide bonds that constitutes the basis for the permanent wave in hairstyling.

---

*Section 2    Major Perming Products*
*Listen to the following audio clip and fill in the blanks with the missing information.*

9.2   Section 2

The major differences of various perming products are determined by the following factors: the 1) _____, 2) _____ of the product and the 3) _____. Most of the waving and neutralizing products for commercial use are 4) _____ solutions, but they also come in other forms as 5) _____. The pH of perming products varies according to the type of reducing agent used. Thioglycolate-based perming lotions have an alkaline pH, as thioglycolates are only effective at an alkaline pH. But both products can cause damage to the hair cuticle and irritate the scalp. As a result, 6) _____ can be added in these formulations to reduce the potential irritation, as they can adjust the pH to 7) _____, making the perms a lot milder. There are also perming lotions with an 8) _____ pH, which are less irritating but less effective as well. Another product available on the market is 9) _____, which produces heat during the perming procedure because of the chemical reaction between 10) _____ and thioglycolic acid. Thermal waves usually give a more pleasant feel on the scalp.

---

**Notes**

Thioglycolates, the salts of thioglycolic acid, are used extensively in cosmetic applications. A majority of thioglycolates, especially ammonium thioglycolates, are employed in the manufacturing of hair perming and hair removal products. A solution comprising ammonium thioglycolate contains ample ammonia, which helps to swell the hair and render it permeable. On the other hand, calcium thioglycolate and potassium thioglycolate are used as ingredients in chemical depilatories to remove unwanted hair from human body.

9.2  Section 3

### Section 3  Different Hair Coloring Products

*Listen to the audio clip and choose the correct answer to each of the following questions.*

1. Why can temporary hair dyes be easily washed out?

   A) Because they contain relatively small molecules.

   B) Because they provide coating on the scalp rather than on the hair cuticle.

   C) Because the binding forces between hair cuticle and the dye molecules are low.

2. Why can semi-permanent hair dyes remain on the healthy hair through 6–8 washing?

   A) Because the dyes are small enough to penetrate the hair cuticle.

   B) Because the dyes can sustain the color from the outside.

   C) Both of the above.

3. How do progressive dye products gradually change the color of hair?

   A) Color is added through water-soluble metal salts which could be deposited on the hair shaft.

   B) Color is added through higher levels of alkalizers.

   C) Color is added by using ammoniacal alkaline agent.

4. What is the difference between semi-permanent and demi-permanent hair dyes?

   A) Semi-permanent hair dyes are more compatible with other hair treatments.

   B) Demi-permanent hair dyes are relatively safer.

   C) Demi-permanent hair dyes can last longer.

5. Why are permanent hair coloring products the most popular today?

   A) Because they come in a wide variety of shades and can both lighten and darken the hair.

   B) Because they produce less allergic irritation and are favored by customers.

   C) Because they make hair styling easier and more fashionable.

9.2  Section 4

### Section 4  Hair Bleaches and Touch-up

*Listen to the audio clip and tell if the following statements are true (T) or false (F).*

1. Hair lightening, or bleaching, is a chemical process that removes the natural or artificial color from the hair.

2. The removal of hair color in bleaching is reversible.

3. The bleach first oxidizes the melanin molecule, and then removes it through chemical reaction.

4. The odor of hair bleaching comes from sulfur released when hydrogen peroxide breaks chemical bonds in the hair.

5. Redyeing is necessary every 6 to 8 weeks.

6. UV light and water exposure can make hair dye colors fade more quickly.

**Notes**

Hydrogen peroxide's chemical description is $H_2O_2$. In lower concentrations, it works well as a disinfectant and antiseptic. When diluted, it can be used to clean and whiten teeth. A different use for hydrogen peroxide is in the creation of beauty products, some of which include hair dyes and bleaching treatments. It may also be added to antibacterial creams and lotions, anti-aging treatments and other facial products.

## Further Listening

9.2　Further Listening

***Thermal Styling***

*Listen to the audio clip and choose the best answer to each of the following questions.*

### Words and Expressions

| | | |
|---|---|---|
| styling *n.* 造型 | hair straightener 拉直板 | curling tongs 卷发钳 |
| bleach *v.* 漂白 | coiled *adj.* 卷曲的 | keratin *n.* 角蛋白 |
| building blocks 基础 | chemical bonds 化学键 | disulfide *n.* 二硫化物 |
| tress *n.* 一绺 | intact *adj.* 完整无缺的 | styling plate 造型板夹 |

1. Which is the best word to describe the thermal styling business in Britain?
   A) Mature.
   B) Booming.
   C) Gloomy.

2. What is the common component for all different types of hair?
   A) Keratin.
   B) Cuticle.
   C) Shaft.

3. Which chemical bond in the hair is reversible?
   A) Metallic bond.
   B) Disulfide bond.
   C) Hydrogen bond.

4. What does the experiment of putting two tresses of hair styled at 185°C and 220°C respectively intend to show?
   A) Hair structure does not change at higher temperatures.
   B) Excessive heat breaks the disulfide bonds and therefore weakens the hair.
   C) The perming effect of thermal styling is irreversible.

5. Which of the following statements is TRUE about using heat on wet hair?

A) It directly breaks the hydrogen bond.

B) It lowers the temperature safe for thermal styling.

C) The maximum heat allowed is under legal regulation.

## Speaking

### Section 1　How to Make Your Hair Dye Eco-friendly?

*Read the paragraph below and interpret it into Chinese.*

Hair dye no doubt makes you look smart and young. Most hair dyes on the market are full of chemicals such as ammonia and PPD, which are linked to immune, nervous system problems and skin irritation, and ammonia causes throat irritation and respiratory problems. When you wash off the chemical-laden hair dye, these dyes get into the water and lead to water contamination. The only hair dye which is completely herbal is natural henna powder without any chemicals at all. The drawback of henna is that it can dry your hair if you do not take care, and it takes a couple of hours to impart color. However, some hair dyes do use chemicals which are not as irritating or environmentally polluting as ammonia or PPD. These non-toxic dyes are available in salons as well as in boxed form. While buying a hair dye product, be sure to read the ingredients and avoid the ones that have a number of chemical compounds.

### Section 2　Think and Discuss

*Suppose you are at a beauty salon for hair styling services. Role-play as customer and stylist and make a conversation. Your talk may include:*

1. Desired hair styles, including the length, waves, colors, etc.
2. Procedures in hair styling and care tips after styling

## Critical Thinking

### Project 1　Motivations for Hair Coloring

Colored hair has become a common statement of identity and fashion. Today, according to some estimates, more than 60% of women in the US color their hair, as do a growing number of men. This market is expected to continue to grow. What is the situation in China? What are the motivators for hair coloring among different groups of people? Carry out a survey and present your findings.

### Project 2　Hair Styling and its Influence on Self-image

Hair is personal because it is a part of our body, yet it is also public because it is on display for others to see. We are judged based on its color, length, and texture. We may believe having the ideal style will help us receive love, wealth, happiness, or achieve a higher social standing. Since people tend to use visual cues to make rapid and often unconscious judgments about the world around them, our hair, along with our clothing and accessories are used to make judgments about us. How does hairstyle affect the way people view us? How does it shape our identity and build our self-image?

## Glossary

1. acid /ˈæsɪd/ *adj.* 酸性的
2. alkaline /ˈælkəlaɪn/ *adj.* 碱性的
3. alkalizer *n.* /ˈælkəlaɪzə(r)/ *n.* 碱化剂
4. ammonia /əˈməʊniə/ *n.* 氨
5. bleaching /ˈbliːtʃɪŋ/ *n.* 漂染
6. cortex /ˈkɔːteks/ *n.* 皮层
7. cysteine /ˈsɪstin/ *n.* 半胱氨酸
8. cystine /ˈsɪstiːn/ *n.* 胱氨酸
9. demi-permanent hair dye 准永久染发剂
10. disulfide bond 二硫键，双硫键
11. fixing lotion 定型剂
12. hair dye 染发剂
13. henna /ˈhenə/ *n.* 海娜，散沫花染剂
14. hydrogen peroxide *n.* 过氧化氢
15. melanin /ˈmelənɪn/ *n.* 黑色素
16. neutral /ˈnjuːtrəl/ *adj.* 中性的
17. neutralization /ˌnjutrələˈzeɪʃən/ *n.* 中和
18. oxidizing agent *n.* 氧化剂
19. perm /pɜːm/ *n. v.* 烫发，卷发
20. perming lotion 烫发剂，软化剂
21. PPD (phenylenediamine /ˈfenəliːnˈdaɪəmɪn/ ) 苯二胺
22. progressive hair dye 渐进染发剂
23. reducing agent 还原剂
24. reduction /rɪˈdʌkʃn/ *n.* 还原
25. rod /rɒd/ *n.* 烫发棒
26. semi-permanent hair dye 半永久染发剂
27. spiral perm 螺旋烫
28. spot perm 局部烫
29. stack perm 层叠烫
30. thermal wave 热烫
31. thioglycolate /θaɪəʊˈglaɪkəleɪt/ *n.* 硫基乙酸盐（酯）
32. touch-up /ˈtʌtʃʌp/ *n.* 补染
33. tress /tres/ *n.* 一绺头发

9.2　Keys and Scripts

## Lesson 3  Solutions for Skin Lightening

### Checklist for Students

**Key Concepts**

| | | | | |
|---|---|---|---|---|
| hyperpigmentation | chloasma | melasma | pregnancy masking | melanin |
| hydroquinone | tyrosinase | tretinoin | arbutin | AHA and BHA |
| laser removal | IPL treatments | chemical peel | | |

**Learning Objectives**

- Understand and explain the popularity of skin-lightening products in the cosmetic industry.
- List different types of ingredients for skin lightening and explain their properties, features and functions.
- Understand the pros and cons of laser treatment for skin lightening.
- Learn the definition of hyperpigmentation and the side effects of skin whitening.

### Before Listening

Depigmentation and skin lightening products, which have been in use for ages in Asian countries where skin whiteness is a major aesthetic criterion, are now also highly valued by Western populations, who expose themselves excessively to the sun and develop skin spots as a consequence. Several modalities of treatment for these problems are available including chemical agents or physical therapies. However, it is recommended that one should be cautious in using skin lightening products as they may contain some ingredients that put people's health at risk.

### Listening

*Section 1   Skin Lightening and the Industry Hype*
*Listen to the audio clip and tell if the following statements are true (T) or false (F).*

9.3   Section 1

1. The pursuit of fair skin has become popular only in recent years.
2. People in sub-Saharan Africa make chemical compounds from ores to lighten their skin.
3. Girls in Rwanda intentionally depigment their skin tones with herbal recipe for special occasions such as marriage.

4. The study carried by Chanel in China suggests that Asian women associate purity of complexion with good health.

5. The white population are more interested in tanning than skin lightening.

6. Melasma is a skin condition caused by pregnancy, thyroid disorders and certain drug treatments.

**Notes**

Women who have medium to dark skin tones are most likely to develop melasma. When melasma appears, it can cause tan, brown, grayish brown, or bluish gray patches and freckle-like spots. These usually appear on certain areas of the face like the cheeks, forehead, chin, and even above the upper lip. While less common, melasma can develop on the arms, neck, or elsewhere. Treatments such as particular creams can help fade the discoloration, but this skin condition may come back.

*Section 2    Major Skin Lightening Ingredients*
*Listen to the audio clip and complete the following chart.*

9.3    Section 2

📖 **Overview**

*Depigmentation can be achieved by (i) regulating the transcription and activity of tyrosinase, (ii) regulating the uptake and distribution of melanosomes in recipient keratinocytes and (iii) interference with melanosomes maturation and transfer. However, as a result of the key role played by tyrosinase in the melanin biosynthesis, most whitening agents act specially to reduce the function of enzyme. Therefore, tyrosinase inhibitors have become increasingly important in the cosmetic and medicinal products used in the prevention of hyperpigmentation and skin whitening.*

| Agents | Mechanisms | Advantages | Disadvantages |
|---|---|---|---|
| Mercury | inactivates tyrosinase enzyme | / | has toxicity, results in darker skin and nails |
| Hydroquinone | 1. | / | 2. |
| Corticosteroids | 3. | / | adrenal suppression, Cushing's syndrome, deteriorated infection, dermatitis, acne, skin thinning, etc. |

(continued)

| Agents | Mechanisms | Advantages | Disadvantages |
|---|---|---|---|
| Ascorbic acid (Vitamin C) | reduces o-dopaquinone to dopa, which prevents melanin formation; also used as antioxidants | 4. | 5. |
| Tretinoin | 6. | / | 7. |
| Azelaic acid | reverses inhibitor of tyrosinase | effective for hyper melanosis and other skin problems that involve proliferation of melanocytes | 8. |

**Notes**

Tyrosinase is any of a family of copper-containing enzymes found in animal and plant tissues, fungi, and bacteria that catalyze the oxidation of phenolic compounds. It is responsible for production of the pigment melanin from tyrosine. Tyrosinase converts a protein building block (amino acid) called tyrosine to another compound called dopaquinone. A series of additional chemical reactions convert dopaquinone to melanin in the skin, hair follicles, the colored part of the eye (the iris), and the retina.

### Section 3  Laser and Light Treatment

*Listen to the audio clip and fill in the blanks with the missing information.*

9.3   Section 3

Topical agents do not work well on pigment in the 1) _____. For example, 2) _____ can relieve pigmentation, but dark spots may reappear quickly. 3) _____, or IPL, and laser treatments can be a much more competitive alternative in this sense. Laser light refers to light amplification by stimulated emission of radiation, which is emitted in a 4) _____ _____. When absorbed by water, hemoglobin, and melanin in the skin, the energy 5) _____. The depth of laser penetration and the amount of melanin to be targeted are determined by the 6) _____. Patients may experience some side effects after laser treatment, such as 7) _____ _____.

In addition, it should be noted that despite the significant effects in skin lightening, IPL and laser treatments can be rather costly with inconsistent results. The discolorations, or brown spots, always reappear if the patients do not apply 8) _____ strictly.

In conclusion, laser and light treatments are proved to have definite effects on pigmentation, making significant change in the appearance in many body parts that are more prone to hyperpigmentation, such as the 9) _____. But consultation of a 10) _____ is always recommended for professional opinion.

## Section 4    Dangers and Side Effects of Skin Whitening

*Listen to the audio clip and choose the correct answer to each of the following questions.*

9.3    Section 4

1. According to the audio clip, which group(s) of women increasingly fall victim to illegal skin whitening products without prescription?
   A) White women.
   B) Asian women.
   C) Hispanic and African women.
2. Which of the following is NOT a side effect of improper use of skin whitening products?
   A) Severe drying, cracking of the skin and itching.
   B) Mercury poisoning, fetal toxicity in pregnant women and Cushing's syndrome.
   C) Heart failure and skin cancer.
3. What is recommended for higher safety of skin lightening solutions?
   A) The authorities must control the use of skin lightening ingredients and raise people's awareness of the potential risks in using such products.
   B) Research is necessary to fully study the effect of temperature on the stability and use of the whitening agents.
   C) Both of the above.

## Further Listening

### Why Do We Have Freckles?

*Listen to the audio clip and complete the following sentences.*

9.3    Further Listening

## Words and Expressions

| | | |
|---|---|---|
| constellation *n.* 星座 | melanin *n.* 黑色素 | melanocytes *n.* 黑色素细胞 |
| clump *n.* 一块，块状物 | lentigines *n.* 痣 | eumelanin *n.* 真黑素 |
| pheomelanin *n.* 棕黑素 | recessive trait 隐性特征 | dominant trait 显性特征 |

1. Freckles are small areas of the skin that contain higher levels of _____.
   A) melanin                     B) glycerin                     C) hormone

2. _____ is the main function of melanocytes.

   A) Acting as soothing agent to allergic skin

   B) Acting as sunscreen that darkens the skin

   C) Acting as moisturizing ingredient that softens the skin

3. Lentigines are different from freckles in that _____.

   A) they contain less melanocytes and therefore fade over time

   B) they contain more melanocytes and therefore do not change

   C) they are caused by excessive exposure to the sun

4. If you have an active MC1R gene, your body produces more eumelanin, which leads to _____.

   A) darker hair and skin that protect yourself against the sun

   B) fairer skin, blonde or red hair and a propensity for freckles

   C) sensitive skin more prone to be damaged by the sun

5. If your MC1R gene is inactive, you will produce more pheomelanin, which leads to _____.

   A) darker hair and skin that protect yourself against the sun

   B) fairer skin, blonde or red hair and a propensity for freckles

   C) sensitive skin more prone to be damaged by the sun

## Speaking

### Section 1  Hyperpigmentation

*Read the paragraph below and interpret into Chinese.*

Hyperpigmentation is a common condition in which some patches of skin turn darker in color. This is a harmless condition caused when there is too much brown pigment, called melanin in the skin. This condition can affect people in all races. Age spots, sometimes called liver spots, are a form of hyperpigmentation. They usually occur because of damage to the skin from the sun. Doctors call these spots solar lentigines. The small, dark spots are found generally on the hands and face, but any area exposed to a lot of sun can be affected, too. There are two types of spots that are similar to age spots, but they cover larger areas of skin. These are referred to as melasma or chloasma spots and, while they are similar to age spots, they are a result of hormonal changes. There are a variety of prescription creams that can help to lighten the darkened patches of skin. They do this by slowing melanin production so that the patches fade. Laser treatments are also effective at removing hyperpigmentation and often can remove the pigmented areas without leaving any scars.

### Section 2  Think and Discuss

*Work in a group. Discuss the following questions and share your answers.*

1. Why are people, whatever their skin tone is, so keen on skin lightening?

2. How is complexion related to personal or social aesthetics, or understanding of beauty?

## Critical Thinking

*Project 1　Pursuit of Fair Skin and Racial Discrimination*

L'Oréal, the world's largest cosmetic and beauty company, announced earlier that it would stop using words like "whitening" and "fair" in describing its products. The announcement came amid anti-racism protests and calls against racial inequality. Unilever has also announced to stop using "fair" on its brand "Fair & Lovely", for the concerns of racial diversity. What do you think of this phenomenon? Do marketing strategies and product descriptions influence social opinions?

*Project 2　Pseudoscience in Skin Lightening Solutions*

A bright, and beautiful complexion would never come away from fashion. There seems to be a plethora of skin-brightening therapies on the market, such as whitening soaps, creams, pills, and even injections. Some people try to use potatoes and honey to achieve skin lightening effect, and some others believe including certain types of herbal foods in the diet would help. Which of the popular solutions are actually "pseudoscience" in skin whitening? What harms do they have?

## Glossary

1. age spot 老年斑
2. ascorbic acid *n.* 抗坏血酸，维生素C
3. blanching /'blæntʃɪŋ/ *n.* 变白，漂白
4. blood vessel *n.* 血管
5. chemical peel 化学剥脱
6. complexion /kəm'plekʃn/ *n.* 肤色
7. contact dermatitis 接触性皮炎
8. corticosteroids /ˌkɔrtɪkəʊ'stɛˌrɔɪdz/ *n.* 皮质类固醇
9. Cushing's syndrome 库欣综合征
10. depigmentation /diːˌpɪgmən'teɪʃən/ *n.* 色素脱失
11. deterioration /dɪ'tɪrɪəreɪt/ *n.* 恶化
12. dicarboxylic acid 二羧酸
13. dopa (dihydroxyphenylalanine) 多巴，二羟基苯丙氨酸

14. dopaquinone /dəʊpeɪkwɪ'nəʊn/ *n.* 多巴醌
15. epidermal turnover *n.* 表皮更新
16. epidermis /ˌepɪ'dɜːmɪs/ *n.* 表皮
17. erythema /ˌerɪ'θiːmə/ *n.* 红斑，红皮病
18. fetal toxicity 胎儿毒性
19. freckle /'frekl/ *n.* 雀斑
20. hair follicle 毛囊
21. hemoglobin /ˌhiːməʊ'gləʊbɪn/ *n.* 血红蛋白
22. Hispanic /hɪ'spænɪk/ *adj.* 西班牙的，拉丁裔的
23. hydrocortisone /ˌhaɪdrə'kɔːtɪzəʊn/ *n.* 氢化可的松，皮质醇
24. hydroquinone /haɪdrəkwɪ'nəʊn/ *n.* 对苯二酚，氢醌
25. hydroxy phenolic compound 酚羟基化合物

26. hyper melanosis 色素沉着过度

27. inactivate /ɪnˈæktɪveɪt/ *v.* 使不活跃，破坏活性

28. inhibitor /ɪnˈhɪbɪtər/ *n.* 抑制剂

29. irradiation /ɪˌreɪdiˈeɪʃn/ *n.* 放射，照射

30. irreversible /ˌɪrɪˈvɜːsəbl/ *adj.* 不可逆转的，无法复原的

31. itching /ˈɪtʃɪŋ/ *n.* 发痒

32. keratinocyte /ˈkerətɪnəsaɪt/ *n.* 角化细胞，角质形成细胞

33. laser /ˈleɪzə(r)/ *n.* 激光

34. light amplification 光放大，光增强

35. liquid nitrogen 液氮

36. liver spot 黄褐斑，老人斑

37. melanin /ˈmelənɪn/ *n.* 黑色素

38. melanocyte /ˈmelənəˌsaɪt/ *n.* 黑色素细胞

39. melanosome /ˈmelənəʊsəʊm/ *n.* 黑素体

40. melasma /məˈlæzmə/ *n.* 黑斑病

41. mercury /ˈmɜːkjəri/ *n.* 汞，水银

42. metal ion 金属离子

43. perioral dermatitis 口周皮炎

44. pigment granule 色素颗粒，色素粒

45. pigment transfer 色素转移

46. prescription /prɪˈskrɪpʃn/ *n.* 处方

47. proliferation /prəˌlɪfəˈreɪʃn/ *n.* 扩散，激增

48. pulsed light device (IPL) 强脉冲光，脉冲强光

49. skin pigmentation *n.* 皮肤色素沉着

50. sun spot 晒斑

51. therapeutic /ˌθerəˈpjuːtɪk/ *adj.* 有疗效的

52. thyroid disorder 甲状腺异常

53. topical agent 局部药剂

54. toxic /ˈtɒksɪk/ *adj.* 有毒的，引起中毒的

55. tyrosinase enzyme 络氨酸酶

56. wavelength /ˈweɪvleŋθ/ *n.* 波长

9.3 Keys and Scripts

# Unit 10　Brand Culture

## Checklist for Students

### Key Concepts

| | | | |
|---|---|---|---|
| cosmetic packaging | material | design | brand recognition and loyalty |
| eco-friendly packaging | PCR | bioplastic | sugarcane　plastic footprint |

### Learning Objectives

• Understand the importance of cosmetic packaging.
• Know what basic information should be on the packaging.
• Learn the tips to make cosmetic packaging stand out.
• Learn some basics about PCR packaging.

## Before Listening

Packaging speaks about the products. A convenient packaging design is one way to communicate to the customers what they can expect from the product. It is also a great way to highlight brand ethos that will help the cosmetic products stand out. Like that of other products, packaging of cosmetics functions to protect, inform, and market. It must be fully compliant with safety requirements in order to ensure an appropriate level of protection of human health. Meanwhile, there is a more outstanding obligation in the beauty and cosmetic companies, that is to woo customers. However, in the modern era, a smart, fancy packaging design is not enough to spread brand awareness and build customer loyalty. With consumers worldwide getting increasingly eco-conscious, sustainability has become a top concern. Beauty businesses must seriously consider going for "green" packaging to promote environmental sustainability.

## Listening

### Section 1　Why Is Cosmetic Packaging Design Important?
*Listen to the audio clip and fill in the blanks with the missing information.*

10.1　Section 1

1) _____. The allure of this industry lies not only in the product itself but also in the packaging. From the shape of the container to overall appearance and presentation, the products are judged from top to toe. The packaging of a product not only plays a significant role in how well the product would be accepted by the consumers, but adds to the growth and success of a business as well.

At its most basic level, product packaging is designed to protect the product within and keep it safe from tampering. There is always the risk of a product arriving damaged while being shipped, or even in rare cases, the vehicle could be involved in a road accident. To reduce this possibility, 2) _____. Also, as cosmetic products are often applied close to the eyes, nose, and mouth, tampering with the product in any way could pose significant health risks to the consumer. A PET tamper-evident locking sleeve, for example, helps keep the product inside safe while also alerting the consumer if the product has been opened or tampered with in any way.

Most of the time the failure of the product is the result of an unattractive or defective packaging because a good percent of the consumers tends to judge the product by its package, or even its color. 3) _____. Foundation and concealer must match the consumer's skin tone exactly, while blush, eye shadow, and lipstick must appear vivid and highly pigmented. Attractive product packaging will attract customers and help them arrive at a decision quickly.

4) _____. As a result, product packaging should be designed in a way that makes it instantly recognizable. A good packaging in an attractive design with the right placement of the logo and a good description of the cosmetic product helps in creating a lasting identity of the brand in the minds of people. When customers view well-designed, high-quality product packaging, they are more likely to put their faith in the company and product and are more likely to make a purchase.

So literally, the entire marketing endeavor depends on the packaging of the cosmetic product. It is the packaging design that 5) _____. Buying the product is a sure thing once their expectations are met.

### Notes

Pentawards, the leading global platform and community for packaging design, is recognized as the world's most prestigious packaging design competition. Among the categories it includes every year, there is one for body, health and beauty, in which the packaging design of various beauty products are judged. The Chinese skincare brand "Herborist" won the silver award in 2008 in the category of luxury cosmetics.

10.1    Section 2

### Section 2    *Cosmetic Packaging Design: What to Include?*

*Listen to the audio clip and tell if the following statements are true (T) or false (F).*

1. "Manufactured From" is used to show the address of the corporate.
2. The young generation are increasingly caring about the design of cosmetic packaging.
3. The most expensive ingredient should be placed at the top of the ingredient list.
4. Only certain products need to have warnings or cautions on the package.
5. Eye makeups usually have shorter shelf lives than other facial cosmetics.

---

**Notes**

The Principal Display Panel (PDP), as it applies to cosmetics in package form, refers to the part of a label that is most likely to be displayed or examined under customary conditions of display for retail sale. The PDP shall be large enough to accommodate all the mandatory label information required to be placed thereon by this part with clarity and conspicuousness. There are specific requirements for the form of stating the required information, including panel display, panel size, style and size of letters, background contrast, and obscuring designs.

---

### Section 3    *Tips for Making Your Makeup Packaging Design Stand Out*

*Listen to the audio clip and fill in the blanks with the missing information.*

10.1    Section 3

In the beauty industry, product packaging is one of the most important tools in a company's arsenal. It's the first thing that catches a customer's eye on the shelf, and it needs to be able to 1) _____ at a glance.

So, how can you make your makeup packaging design stand out from the crowd?

2) _____

Let's be honest, we've all been guilty of 3) _____ because the packaging was pretty. But unless you're a makeup artist, you don't need a ton of different colors and products.

In fact, you can create any look with just a few basic items. And when it comes to packaging, less is definitely more. A 4) _____ or 5) _____ _____ will never go out of style.

Plus, it's easy to find what you're looking for in your bag or medicine cabinet. So save yourself some money and streamline your beauty routine by keeping your makeup packaging simple.

6) _____

Your packaging is a reflection of your brand, so you want to make sure that it's made with high-quality materials. Use 7) _____ for your box and choose a printing process that will give your design 8) _____.

A professional printer, such as CustomBoxesWorld, can help you create a beautiful, eye-catching design that will make your brand stand out from the competition. Investing in high-quality packaging is an important part of building a successful business.

9) _____

Makeup packaging comes in all sorts of shapes and sizes, so there are no hard and fast rules when it comes to choosing a shape for your own product.

The most important thing is to be creative and think outside the box—10) _____ _____ —to come up with a shape that will really make your product pop. Get inspired by the world around you, and have fun with it. You might be surprised at just how much of a difference the right packaging can make.

### Section 4   Eco-friendly Packaging
*Listen to the audio clip and give short answers to the questions below.*

10.1   Section 4

1. Why should cosmetic companies employ sustainable packaging?
_____

2. What does PCR stand for?
_____

3. What is virgin resin made from?
_____

4. How to make virgin plastic packaging into recycled packaging?
_____
_____

5. What are the advantages of PCR packaging?
_____

## Further Listening

10.1   Further Listening

### Cosmetic Packaging Innovations
*Listen to the audio clip and give short answers to the questions below.*

## Words and Expressions

refillable *adj.* 可再装的                    biodegradable *adj.* 可生物降解的
landfill *n.* 垃圾填埋场                     PET 聚对苯二甲酸乙二醇酯
LDPE 聚乙烯                              HDPE 高密度聚乙烯
PP 聚丙烯                                sugarcane *n.* 甘蔗
troll *n.*（钓鱼）拽绳                        opaque *adj.* 不透明的

1. How many new packaging materials are introduced? What are they?

_____

2. How long does it take for normally plastic packaging to degrade? How about the biodegradable packaging?

_____

3. Why does the speaker say the sugarcane bioplastic is good for presenting the product?

_____

4. What is the difference between ocean waste plastics and other materials?

_____

## Speaking

### Section 1  *A Good Example of Packaging Design: Benefit*
*Read the paragraph below and interpret it into Chinese.*

  *Benefit* is a funky cosmetic brand whose design is inspired by vintage influences. Pinup girls line most of the products, cute and intricate drawings of beautiful, happy-go-lucky women straight out of the 1920s. This imagery is extremely engaging and adds a playful edge to the design as a whole. This brand also takes advantage of bright and exciting colors. Pinks, whites and silvers are the main colors used here, either working as a background color, as an accent or in the typography. This cosmetic packaging design really pops because of the imagery, vintage vibes and enthusiastic use of color. It is unique, calming and beautiful and they keep this theme consistent across all of their packaging. It's obviously much easier to tell a story on a tin or box than it is on a tube of mascara, but *Benefit* is able to weave this story and these characters across all of its product offerings to align itself as a brand that is fun, cool and in touch with the times.

### Section 2  *Think and Discuss*
*Work in a group. Discuss the following questions and share your answers.*

1. Though we see many fancy cosmetics packaging every year, some of them are time-tested and classic. Do you find the packaging of any product or line of product particularly amazing? Why?
2. Packing of cosmetic products pose great threat to the environment. Are there ways to enjoy cosmetics and personal care products while reducing one's plastic footprint?

## Critical Thinking

### Project 1  *The Ugly Side of Beauty*

  Promising to make us more beautiful, cosmetic industry is actually creating a lot of ugly waste in the name of beauty. How many units of packaging are produced every year and how many of them are truly recycled? Why are cosmetic companies reluctant to develop sustainable plans? Are there any pioneer companies that take a more productive approach to the packaging issue? Study these questions and present your findings.

*Project 2    Bad Packaging!*

If good packaging design makes the beauty products stand out and increases customers' loyalty, bad packages can do just the opposite. More than once we see how some beauty brands disappoint their customers with ugly designs, excessive or even deceptive packages or clumsy containers. Please conduct a study on the cases of terrible cosmetic packaging and give suggestions on what improvement can be made.

## Glossary

1. Biodegradable *adj.* 可生物降解的
2. degradation /ˌdegrə'deɪʃn/ *n.* 降解
3. detergent /dɪ't3:dʒənt/ *n.* 洗涤剂，清洁剂
4. High Density Poly Ethylene (HDPE) 高密度聚乙烯
5. Low Density Poly Ethylene (LDPE) 低密度聚乙烯
6. polymerase chain reaction (PCR) 聚合酶链式反应
7. polyethylene terephthalate PET 聚对苯二甲酸乙二醇酯
8. petrochemical /ˌpetrəʊ'kemɪkl/ *n.* 石油化学产品
9. polycondensation /ˌpɒlɪˌkɒnden'seɪʃən/ *n.*
　　缩聚（作用）
10. polymerization /ˌpɒlɪməraɪ'zeɪʃn/ *n.* 聚合作用
11. polypropylene /ˌpɒli'prəʊpəli:n/(PP) *n.* 聚丙烯
12. resin /'rezɪn/ *n.* 天然橡胶
13. tamper-evident *adj.* 防拆封的

10.1　Keys and Scripts

## Lesson 2  Global Competitors

### Checklist for Students

**Key Concepts**

| | | | |
|---|---|---|---|
| entrepreneurial spirit | marketing strategy | global sale | revenue |
| single channel | department store | visibility | brand recognition |
| brand loyalty | K-pop craze | celeb-centered advertising | C-beauty |
| digital marketing | TCM | kiosk | positioning strategy |
| e-platform | | | |

**Learning Objectives**

● Learn the six core values of L'Oréal company.

● Analyze the reasons for the declining market share of foreign brands in China.

● Summarize the reasons for the popularity of Korean cosmetic brands.

● Examine the rise of Chinese domestic cosmetic brands.

● Reflect on the development strategies of global cosmetics competitors and predict the future market trends.

### Before Listening

The global cosmetics market was valued at USD 307.69 Billion in the year 2020. The huge demand for cosmetics is influenced by awareness among people about the benefits of cosmetics for their skin and hair, which uplifted their average expenditure on cosmetics. Moreover, the increasing trend for sun care products, night skin repair creams, fresh face mists is likely to boost demand for cosmetic products in the future. There is also a growing desire for health-promoting and self-care products. Asia Pacific region is estimated to be the fastest-growing region in cosmetics market, due to growing concern over health and hygiene, personal appeal, and rising demand for natural and organic beauty products and many others. China, Japan, South Korea, India are the major cosmetics markets of the world. The online segment is expected to grow at the fastest rate, and the advantages of e-commerce have benefited not only businesses but also customers in terms of cost and a wide range of possibilities.

## Listening

### *Section 1    Six Values of L'Oréal*
*Listen to the audio clip and fill in the blanks with the missing information.*

10.2    Section 1

Passion, innovation, entrepreneurial spirit, open-mindedness, quest for excellence and responsibility are the guidelines of L'Oréal.

#### Passion

We have passion for what cosmetics can bring to women and men: 1) _____ _____. Passion is also the key for a business which is intrinsically linked to 2) _____ _____, because creating beauty products and services means seeking to understand others, knowing how to listen to them, 3) _____, and anticipating their needs. This passion is what makes the L'Oréal adventure so fascinating.

#### Innovation

Innovation is also one of L'Oréal's founding values. For a company founded by a scientist, innovation is essential because beauty is 4) _____ that constantly requires 5) _____. At L'Oréal, people always want to push back the limits of knowledge and discover new ways to create products and services that are truly different and surprising.

#### Entrepreneurial spirit

As a synonym for 6) _____, entrepreneurial spirit has always been encouraged and embodied in a specific management style. Today it is still the driving force behind L'Oréal built above all on a belief of the importance of each individual and their talents.

#### Open-mindedness

Listening to consumers and understanding their culture, being open to others and benefiting from their differences are absolute priorities in order to respond to the 7) _____ _____ around the world. They are inseparable from our business and our mission.

#### Quest for Excellence

This value permeates every aspect of our business in every country and that is expressed in a state of mind and 8) _____. We all share this desire to surpass ourselves to be able to provide the best for our consumers.

## Responsibility

L'Oréal's first invention, the "9) _____ " was already an expression of this fundamental concern for effective, safe and innocuous products. But our sense of responsibility goes far beyond that. As a world leader in beauty, we have, more than others, the duty to protect the beauty of the planet and to contribute to 10) _____ _____ with which we engage.

> **Notes**
>
> L'Oréal is the third largest cosmetics company in the world. It was founded in 1909 by Eugene Schuler who was a French chemist. It is headquartered in Clichy, Hauts-de-Deine and has a portfolio of more than 500 brands from many cultures. L'Oréal is a multinational conglomerate which has its operational presence in more than 130 countries. The various products they offer are being researched at their 6 worldwide research centers, ranging from skincare to make-ups and all other kinds of cosmetics.

### Section 2   Revlon Kisses China Goodbye
*Listen to the audio clip and give short answers to the questions below.*

10.2   Section 2

1. What challenges do western beauty brands face in maintaining their leadership?

2. What do we know about Revlon's global sales in recent years?

3. According to analysts, why did Revlon fail to prosper?

4. What does the Mintel's latest consumer survey say about Chinese women's preference of color cosmetics?

5. What new products have some international brands developed?

### Section 3   Popularity of Korean Cosmetics
*Listen to the audio clip and choose the correct answer to each of the following questions.*

10.2   Section 3

1. How do Korean cosmetic companies shape an aspirational idea of beauty?

A) Through dermatologist recommendation.

B) Through clever, celeb-centered advertising.

C) Through organic, nonallergic ingredients.

2. Why is Missha able to establish a foothold overseas amid a massive expansion?

A) It offers affordable prices.　　B) It has more chain stores.　　C) It changes local aesthetics.

3. What products does LG focus on promoting in Vietnam?

A) Moisturizers.　　　　　　B) Whitening creams.　　　　C) Sun creams.

4. According to the report, what Korean cosmetic is popular in Hong Kong?

A) Moisturizing masks.　　　B) Hydrating lotions.　　　　C) BB creams.

5. What can be inferred from the case that Dr. Jart became the first Korean cosmetics supplier to Boots, UK's biggest chain of drug store?

A) Korean cosmetics have passed rigid animal testing.

B) Korean cosmetics have high quality.

C) Korean cosmetics specialize in cosmeceuticals.

### Section 4　Rise of Chinese National Cosmetic Brands

*Listen to the audio clip and tell if the following statements are true (T), false (F), or not given (NG).*

10.2　Section 4

1. Large multinational cosmetics brands are pulling out of the Chinese market.

2. Chinese beauty companies become leading players in the cosmetic industries for their digital marketing, social commerce strategies and powerful retail networks.

3. Shanghai Jahwa United Co., Ltd is looking into acquiring foreign brands to break into the international cosmetic market.

4. Herborist positions its skincare products as a combination of traditional Chinese medicine (TCM) and modern cosmetic technology.

5. Chinese consumers prefer chemical ingredients to natural alternatives as they have stronger effects.

6. Proya has made successful nationwide expansion through its smart positioning strategy.

7. Proya aims at high-income customers in China's tier-one and tier-two cities.

8. China's beauty market prospers on e-platforms, which takes a toll on foreign rivals.

## Further Listening

### The Economics of Sephora

*Listen to the audio clip and choose the correct answer to each of the following questions.*

10.2　Further Listening

## Words and Expressions

agnostic *adj.* 不可知论的，怀疑的　　　　　　exclusive *adj.* 专有的，独有的

inclusivity *n.* 包容性

1. How did customers buy cosmetics in the past? How do they do in Sephora?

_____

_____

2. Why do customers have more trust in Sephora's sales representatives?

_____

_____

3. What are the three tiers of Sephora's Beauty Insider program?

_____

_____

4. How to get to the top two tiers in the program?

_____

_____

5. How does the program benefit Sephora?

_____

_____

6. What is Color IQ? Why did Sephora relaunch it?

_____

_____

### Notes

The new Color IQ leverages a cutting-edge proprietary algorithm that provides customers with a dataset of 10K-plus skin tones, suitable across all shade ranges. In fact, the introduction of Artificial Intelligence and Augmented Reality technology to the beauty industry has modernized the capabilities of beauty brands and the experiences they can provide a consumer with. Besides Sephora, companies like Estée Lauder and L'Oréal have also utilized AI and AR technology to launch personalized virtual try-on makeup solutions.

## Speaking

### Section 1   Development Strategies for C-beauty Brands
*Read the paragraph below and interpret it into Chinese.*

C-beauty brands must develop a distinct niche that distinguishes them from competitors. China's cultural capital of C-beauty still lags behind that of regional competitors from South Korea and Japan. South Korea's burgeoning entertainment industry is partly responsible for the popularity of Korean beauty products, while Japanese brands are often associated with high quality. However, C-beauty brands still lack a distinctive core idea. One potential idea could be associating with traditional

Chinese wellness techniques and beauty practices. Global consumers have shown interest in traditional Chinese culture previously. In 2018, for example, the jade roller, a traditional Chinese tool for facial massaging, went viral in the US, and sales increased rapidly. Cultural practices from ancient China, TCM ingredients, and other aspects of China's heritage could help Chinese beauty brands find a distinctive lure. However, brands need to strike a balance between being unique and catering to the preferences of international consumers. This means finding a way to appeal to customers who might not have any connection to Chinese culture or beauty ideals. While Chinese brands should undoubtedly keep their Chinese characteristics, they need to adapt their marketing strategy to appeal to a more global audience.

### *Section 2    Think and Discuss*
*Work in a group. Discuss the following questions and share your answers.*

1. What are the new beauty trends in the cosmetics market? How will the global cosmetics market develop in the future?
2. What can be done to project the image of a cosmetic brand and to enhance its international competitiveness?

## Critical Thinking

### *Project 1    Challenges for Foreign Cosmetic Brands in China*

Foreign cosmetics brands jointly grabbed 57.9% of the mainland Chinese market as of May 2009, but their market share fell to 44.5% as of May 2012. Procter & Gamble and Avon both lost market share in the six years through 2012. At the same time, the rise of confident Chinese and South Korean brands, such as China's Herborist, Chinfie, CMM, Houdy, Caisy and Longrich brands, and South Korean brands AmorePacific, Sulwhasoo and Missha, poses the long-term threat to their profits. Keeping up with fast-changing China is increasingly harder for larger MNCs. What other challenges do foreign cosmetics brands face? How can MNCs cope with them?

### *Project 2    Success of Chinese Domestic Makeup Brands*

For a long time, the Chinese cosmetics market has been dominated by international brands such as L'Oréal, Lancôme and Estée Lauder. However, a great change is now taking place in this market. Perfect Diary was the first makeup brand whose sales reached 100 million CNY in "Double 11 Festival". At the end of 2019, Tencent's "2019 Domestic Beauty Brand Insight Report" showed that Chinese domestic brands accounted for 56% of the Chinese cosmetics and market. What marketing strategies account for the success of these domestic makeup brands?

## Glossary

1. Beiersdorf 拜尔斯道夫
2. Dr. Jart 蒂佳婷
3. Herborist 佰草集
4. Langeige 兰芝
5. Liushen 六神
6. Missha 谜尚
7. Nivea 妮维雅
8. Olay 玉兰油
9. Pond's 旁氏
10. Procter & Gamble 宝洁

11. Proya 珀莱雅
12. Sephora 丝芙兰
13. Shanghai Jahwa 上海家化
14. Watsons 屈臣氏

10.2  Keys and Scripts

# References

1. Ahmed, H. A. (2010). Review on skin whitening agent. *Khartoum Pharmacy Journal*, 13. 5–9. No.1 June. 2010

2. Baki, G. & Alexander, K. S. (2015). *Introduction to cosmetic formulation and technology*. New Jersey: John Wiley & Sons, Inc.

3. Begoun, P. (2009). *The original Beauty Bible* (3rd ed.). Washington: Beginning Press.

4. Benson, H. A. E. (Eds). (2019). *Cosmetic formulation principles and practice*. London: CPC Press.

5. Flynn, T.C. et al (2001). Dry skin and moisturizers. *Clinics in Dermatology,* 19. 387–392.

6. Iwata, H. & Shidama, K. (2013). *Formulas, ingredients and production of cosmetics*. Tokyo: Springer.

7. Latita, G. & Shalini, G. (2020). Creams: A review on classification, preparation methods, evaluation and its applications. *Journal of Drug Delivery and Therapeutics*, 10. 281–289.

8. Li, S. Y. [李思彦] (编). (2020). 化妆品专业英语. 北京：化学工业出版社.

9. Jones, G. (2010). *Beauty imagined a history of the global beauty industry*. Oxford: Oxford University Press.

10. Pageon, H. et al. (2010). Glycation and skin aging. In. Farage, M. A. (Eds). Textbook of Aging Skin (2nd ed.), (pp 1248–1264). Berlin: Springer-Verlag.

11. Rai, S. et al. (2015) Regulations of cosmetic across globe. *Applied Clinical Research Trail & Regulatory Affairs*, 2. 137–144.

12. Shai, A. (Eds). (2009). *Handbook of cosmetic skin care* (2nd ed.). London: Informa Healthcare.

13. Wen, J. (2022). Anti-ageing peptides and proteins for topical applications: A review. *Pharmaceutical Development and Technology*, 27. 108–125.